The Conformational Analysis of Heterocyclic Compounds

The Conformational Analysis of Heterocyclic Compounds

FRANK G. RIDDELL

University of Stirling

1980

ACADEMIC PRESS

A Subsidiary of Harcourt Brace Jovanovich, Publishers

London New York Toronto Sydney San Francisco

ACADEMIC PRESS INC. (LONDON) LTD.
24/28 Oval Road,
London NWI

United States Edition published by
ACADEMIC PRESS INC.
111 Fifth Avenue
New York, New York 10003

British Library Cataloguing in Publication Data

Riddell, Frank G.
   The conformational analysis of heterocyclic
   compounds.
   1. Heterocyclic compounds
   2. Conformational analysis
   I. Title
   547′.59   QD400   79-41514

   ISBN 0-12-588160-6

Filmset by Composition House Ltd.
Salisbury, Wiltshire

Printed in Great Britain by
Whitstable Litho, Whitstable, Kent

Preface

In 1969 in his Nobel Prize lecture, Sir Derek Barton, the co-founder of conformational analysis, said: "The introduction of heteroatoms into a cyclohexane ring as in piperidine or pyran raises conformational problems of considerable interest and sophistication." Throughout the late 1960s and all through the 1970s, Barton's words have been found to be true in intensive research into heterocyclic conformations. This book is an attempt to discuss some of the interesting and sophisticated problems that have arisen from this work.

In a field as large as the conformational analysis of heterocyclic compounds it is clearly impossible to attempt a comprehensive coverage of all published work. Because of this I have tried to present a broad framework of theory in Chapters 1 and 2 on which the discussion of the specific ring systems may be hung in Chapters 3 to 7. This means that I have attempted to concentrate on those aspects of the subject, those experiments and those results, that I regard as fundamental to an understanding of the basic theory, and to show how a limited number of theoretical concepts can explain a wide variety of results in many different systems.

In presenting values of energies I have stuck to kcal mol^{-1}. Although these units are not now part of the modern SI system they are still extensively used in this area, are more widely understood, and will undoutedly continue to be used by chemists for many years. Interatomic distances, however, are all presented in the SI unit of pm.

By far the largest section of the book is concerned with six-membered ring systems. To simplify presentation, the heteroatoms are introduced successively during Chapters 4 to 6. Thus, Chapter 4 deals exclusively with oxygen-containing rings, and Chapter 5 with nitrogen-, and nitrogen- and

v

oxygen-containing rings. Chapter 6 contains rings with oxygen, nitrogen, sulphur and phosphorus.

In preparing this book I have assumed that the reader has a knowledge of organic stereochemistry up to that of a general or honours degree in a British University. The book should therefore be of interest, and I hope of some use, to graduate workers in organic chemistry, be they in academic or industrial positions.

It is a pleasure to thank those of my colleagues who have helped in preparation of this book, in particular, Dr M. J. T. Robinson of Oxford University and Professor E. L. Eliel of the University of N. Carolina for their constructive comments on the draft manuscript, and Drs P. Murray-Rust and H. Maskill in Stirling University for helpful and detailed advice on parts of the manuscript. I would particularly like to thank Mrs J. Weber for her efficient typing of the manuscript. Last but not least, I wish to thank my wife, Joan, for her patience and understanding during the time this book was being prepared, and for helping to check the manuscript.

F. G. Riddell
February 1979

Contents

Preface   v

1 Introduction   1

Historical Definitions—Conformation-directing forces—Acyclic
compounds—Alicyclic compounds

2 Special Conformational Considerations for
Heterocycles   22

Bond lengths, van der Waals radii and non-bonded interactions—
Torsional interactions—Electrostatic interactions—Hydrogen
bonding—Methods of investigation

3 Four- and Five-Membered Rings   47

Four-membered rings—Five-membered rings

4 Six-Membered Oxygen-Containing Rings   66

General considerations—Tetrahydropyran—1,2-Dioxan—
1,3-Dioxan—1,4-Dioxan—1,3,5-Trioxan—1,2,4,5-Tetroxan

5   Six-Membered Nitrogen-Containing Rings   83

General considerations— Piperidine—Hexahydropyridazine—
Hexahydropyrimidine—Piperazine—Hexahydro-1,3,5-triazine—
Hexahydro-1,2,4,5-tetrazine—Tetrahydro-1,2-oxazine—
Tetrahydro-1,3-oxazine—Tetrahydro-1,4-oxazine (morpholine)—
Tetrahydro-1,4,2-dioxazine—Tetrahydrooadiazines—Nitrogen-
containing fused rings

6   Six-Membered Rings Containing Sulphur
     and Phosphorus   104

Sulphur-containing rings—Phosphorus-containing rings

7   Seven-Membered and Larger Rings   131

Introduction—Seven-membered rings—Eight-membered rings—
Nine-membered rings—Ten-membered rings—Larger rings

Index   149

I dedicate this book
to the memory of my parents.

To them I owe all that I became.

Edgar Lee Masters
"The Spoon River Anthology"

1
Introduction

Historical

In 1874, just over one hundred years ago, the foundations of all stereo-chemistry were laid by van't Hoff and Le Bel, who independently proposed that the four valencies of carbon are directed towards the vertices of a regular tetrahedron.[1] Although their ideas were not accepted immediately by all the chemical world their concept of the three-dimensional nature of molecules gradually gained sway.[2] Conformational analysis made its first tentative steps in 1890 when Sachse[3] suggested that cyclohexane could exist in two arrangements, free from angle strain, which were later termed "chair" and "boat" conformations. Sachse's intuition into the nature of cyclohexane was quite remarkable. He was aware that one form of the molecule was rigid and that the other was continuously flexible, and that angle deformation is needed to go from one form to the other. Sachse's ideas lay neglected for years until Mohr revived them in 1918[4] and Hückel found support for them in 1925 by isolating two decalins (*cis* and *trans*).[5] Progress in conformational analysis thereafter slowly gained momentum. In the 1920s and 1930s three-dimensional thinking in sugar chemistry was introduced by Boeseken,[6] Haworth introduced the word "conformation",[7] Bilicke found the chair form of cyclohexane in studies on hexachlorocyclohexanes[8] and Kohlrausch recognized two types of bond in the chair form of cyclohexane (now termed axial and equatorial) from his Raman studies of cyclohexyl halides.[9]

1

In the 1930s Hassel started work on cyclohexane initially making use of dipole moment measurements and X-ray diffraction studies of mono-, di-, and polysubstituted cyclohexyl halides. Dipole moment studies are a blunt tool for conformational studies, which can sometimes give ambiguous results, whilst X-ray crystallography only gives results in the solid phase. It was therefore not until 1938 when Hassel started his electron diffraction work in the gas phase that unambiguous results free of solid phase inter-molecular interactions were forthcoming and more substantial progress was made. Hassel's results revealed cyclohexane derivatives in the gas phase with both equatorial and axial substituents, and soon most of the basic structural facts on which conformational analysis is based were clear to him. Because of the German occupation of Norway during the Second World War and Hassel's imprisonment from 1943 to 1945, most of his significant work appeared in Norwegian in rather obscure journals. However, in 1943 he clearly set forward the basic conformational facts about cyclohexane:[10]

(i)   The chair form is the preferred conformation.
(ii) Substituted cyclohexanes undergo a ring inversion process and the substituent may occupy either the axial or the equatorial position.
(iii) In all cases then known the equatorial position was favoured for the substituent, although in 1943 it was not entirely clear what energy differences were involved.

It was some time after the war before Hassel's contribution began to bear fruit. Barton, in 1950, published a pioneering paper which revolutionized organic chemistry.[11] By building on Hassel's ideas and applying them to the chemistry of natural products he was able to explain many previously obscure results. The greater stability of equatorial over axial substituents, their greater reactivity in reactions subject to steric hindrance and their lesser reactivity in reactions where strain is relieved in the transition state, could now be used as basic stereochemical principles. For example, equilibration conditions will convert an axial alcohol into its equatorial epimer, esterifica-tion of carboxylic acids or hydrolysis of esters, is more rapid for equatorial than for axial substituents, but the reverse holds for chromic acid oxidation of alcohols.

These ideas lead to a new and fundamental understanding of many reactions and have directly and indirectly assisted in many subsequent advances in organic chemistry.

In 1969 Hassel and Barton were jointly awarded the Nobel Prize for chemistry for "developing and applying the principles of conformation in chemistry".[12]

The impetus given to the investigation of conformations by these early workers has continued. Subsequent workers have been concerned with investigating the chair conformation in more detail than was possible by Hassel, in measuring rates of interchange and energy differences between conformations, in expanding the subject from cyclohexane to a wide variety of other cyclic and acyclic systems, and with what concerns us in this book, extending the principles of conformational analysis into heterocyclic compounds.[13] The first steps in this last-mentioned direction were made by Barton and Cookson who in 1956 showed that several features of heterocyclic chemistry could be explained on the basis of conformational theory.[14] The first comprehensive review of heterocyclic conformational analysis in 1967 gave an added boost,[15] and since then the subject has grown so much that even a full book such as this can barely hope to do the subject justice.

Definitions

What is conformational analysis? We may answer this question by stating the fundamental tenet of conformational analysis:[14] ". . . that the chemical and physical properties of organic molecules depend not only on their gross structure and stereochemistry, but also on the conformations they prefer to adopt". What then are conformations? This question is best answered by placing the term "conformation" in the context of stereochemistry as a whole. We can recognize a stereochemical hierarchy consisting of three levels or types of isomerism. The first level differentiates between constitutional isomers which differ in the connectivity of their atoms. The order in which the atoms are joined in ethanol (1) differs from that in dimethyl ether (2). This different sequence of directly bound atoms is a simple, readily apppreciated criterion to distinguish between constitutional isomers.

$$CH_3-CH_2-OH \qquad\qquad CH_3-O-CH_3$$
$$(1) \qquad\qquad\qquad (2)$$

The second level in our hierarchy consists of stereoisomers which have the same sequence of directly bound atoms but which differ in their arrangement in space. *Cis* and *trans* olefins such as the 2-butenes (3 and 4) or *threo* and *erythro* isomers such as threose (5) and erythrose (6) fall into this category. No matter how one transposes these molecules in space the one will never be identical with its stereoisomer. A special type of stereoisomeric relationship

(3)   (4)

is that between an object and its non-superimposable mirror image. This enantiomeric relationship is illustrated by the stereoisomeric and enantiomeric lactic acids (7 and 8). All stereoisomers which are not enantiomers will be called diastereoisomers. Thus (3) and (4) are diastereoisomers as are (5) and (6), but (7) and (8) are enantiomers. All three pairs of compounds differ in their configuration.

$$
\begin{array}{cc}
\mathrm{CHO} & \mathrm{CHO} \\
\mathrm{HO}\!-\!\!|\quad & |\!-\!\mathrm{OH} \\
\quad |\!-\!\mathrm{OH} & |\!-\!\mathrm{OH} \\
\mathrm{CH_2OH} & \mathrm{CH_2OH} \\
\text{D-(}-\text{)-threose} & \text{D-(}-\text{)-erythrose} \\
(5) & (6)
\end{array}
$$

$$
\begin{array}{cc}
\mathrm{CO_2H} & \mathrm{CO_2H} \\
\quad |\!-\!\mathrm{OH} & \mathrm{HO}\!-\!| \\
\mathrm{CH_3} & \mathrm{CH_3} \\
(-)\text{-lactic acid} & (+)\text{-lactic acid} \\
(R\text{-configuration}) & (S\text{-configuration}) \\
(7) & (8)
\end{array}
$$

The third level in the hierarchy consists of conformations. Many people define a conformation as any arrangement of a molecule formed by rotation about single bonds. On this definition ethane or *n*-butane, for example, have an infinite number of conformations. This view is taken up by the IUPAC rules[16] which give the following definitions "(a) Classical interpretation: the conformations of a molecule of defined configuration are the various arrangements of its atoms in space that differ only as after rotation about single bonds; (b) this is usually now extended to include rotation about π bonds or bonds of partial order between one and two; (c) a third view extends the definition further to include also rotation about bonds of any order including double bonds." Later the same tentative rules continue: "A molecule in a conformation into which its atoms return spontaneously after small displacements is termed a conformer."

These definitions unambiguously restrict the use of the words conformation and conformer to arrangements formed solely by rotation about bonds. In many respects these and similar definitions are inadequate. For cyclohexane to go from a chair to a twist conformation requires not only rotation about single bonds but also distortion of bond angles. Since bond angle distortion is not allowed for on this type of definition the chair (9) and twist (10) forms of cyclohexane cannot be conformations—they must be stereoisomers! In piperidine there are, what most people would call, two conformations, which can be interconverted either as in cyclohexane, by a ring

(9)   (10)

inversion via a twist conformation ($\Delta G^{\ddagger} \sim 11$ kcal mol^{-1}), or more easily by inversion at the three coordinate nitrogen atom ($\Delta G^{\ddagger} \sim 6$ kcal mol^{-1}) (Fig. 1.1). Again the latter mode of conformational interconversion is not allowed by the IUPAC rules. However, if the rules are stretched a little to allow ring inversion to be a conformational process one can then say that all the species in Fig. 1.1 are conformations.

This cannot be the case for 3-methylpiperidine! The sequence of ring and nitrogen inversions is shown in Fig. 1.2 which is similar to Fig. 1.1. The two left-hand-side species are conformations and the two right-hand-side species are conformations, but under the IUPAC rules the left-hand-side pair and

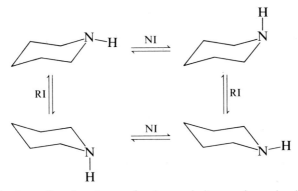

Fig. 1.1   Conformational route map for ring and nitrogen inversion in piperidine.

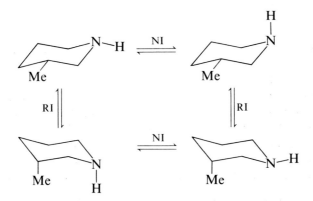

Fig. 1.2   Conformational route map for ring and nitrogen inversion in 3-methylpiperidine.

the right-hand-side pair are related as stereoisomers. The common and widespread usage of conformation and conformer would call all the species in Fig. 1.2 conformations of the same compound. Quite clearly there is more to defining conformation than a bald statement about bond rotation.

From the above discussion we can see that conformations are in fact a type of stereoisomer. Conformations differ in the spatial arrangement of their atoms but not in their bonding sequence. Our definition must therefore include this idea. For the purposes of this book we shall adopt the following definition. Conformations are stereoisomers that can be interconverted either by rotation about bonds of order approximately one, with any concomitant small distortions of bond lengths and angles, or by inversion at a three-coordinate centre in the molecule, or by pseudorotation on phosphorus.

One final point needs dealing with. The bulk of experimental results on the structure, and therefore the conformations of molecules, comes from observations on molecules in or near their ground states i.e. occupying a potential-energy minimum. The above definition allows us to treat conformations in the same light as stereoisomers—as an assembly of molecules occupying a potential-energy minimum. Those forms of molecules such as eclipsed ethane which do not have a finite existence, and in fact correspond to transition states, do not fall under our definition. They will, when we have to discuss them, be called arrangements.*

Other words have appeared in the literature meaning largely the same thing as conformation or conformer. "Rotational isomer" first appeared in German (*Rotationsisomere*) and now sometimes occurs as rotamer.† The word "constellation" has also been used in the German literature.

Conformation-Directing Forces

In principle the preferred conformations of any molecule can be determined by a complete molecular orbital study. In practice this is an extremely expensive exercise in computing and is not possible for all but the simplest of cases. It is common for workers in this field to adopt a series of approximations to lighten the computing burden. Complete Neglect of Differential Overlap (CNDO), Intermediate Neglect of Differential Overlap (INDO) and Modified Intermediate Neglect of Differential Overlap (MINDO) have all been suggested and argued for. In fact, although remarkably good results are claimed and obtained for these approaches, each invariably leads to a different picture of intramolecular forces from that provided by the

* This distinction has been made from time to time in the past. For a discussion of this point and leading references see Ref. 13b.
† Rotamers are a subset of conformers.

others. If the experts in this field cannot agree on a common approach, who am I to recommend one to you!

Chemists and molecular physicists generally, therefore, turn to a "classical" picture of interactions inside a molecule. The problem is broken down into a series of smaller components each with a readily appreciated physical basis. This results in "molecular mechanics" calculations which are certainly much easier to handle computationally than any complete MO study, especially for larger molecules. For a typical modern molecular mechanics force field the reader should consult Ref. 17.

In this classical picture of the forces that control conformational preferences the following headings are most commonly used:

1. Torsional forces
2. Non-bonded interactions
3. Bond stretching
4. Bond angle deformation
5. Electrostatic interactions
6. Solvent effects
7. Hydrogen bonding
8. Resonance effects

Considerable effort has been put into devising functions which describe the way molecular energy varies with headings 1–4. Although functions are available for 5 they are mainly useful for qualitative rather than quantitative interpretations. Progress is being made on the evaluation of solvent effects on conformational equilibria. However, attempts at estimations of 7 and 8 are at present strictly enlightened guesses, although MO calculations on hydrogen bonding have been successful.

The discussion which follows is an attempt to give a general picture of the way in which molecular energy changes with geometry. In general the terms and equations used in one force field are not transposable into another force field.

Torsional forces

The general form of the potential function for rotation about a single bond is given by the Fourier expansion

$$E(\phi) = \tfrac{1}{2} \sum_n E_n(1 + \cos n\phi) \qquad \text{eq. 1.1}$$

where $E(\phi)$ is the energy of the arrangement in which there has been a rotation of ϕ from the eclipsed position. The terms E_n in the expansion are the energies associated with each n-fold component of the total function.

To take a simple case, ethane, with three-fold symmetry along the carbon–carbon bond, only terms involving $\cos 3\phi$ are significantly involved, and the expression becomes

$$E(\phi) = \tfrac{1}{2}E_0(1 + \cos 3\phi) \qquad \text{eq. 1.2}$$

where E_0 is found experimentally[18a] to be 2.928 kcal mol^{-1} (Fig. 1.3).

In the more complicated case of *n*-butane where there are two distinct types of barrier, a one-fold component, associated with eclipsing of the methyl groups, and a three-fold barrier the expansion becomes

$$E(\phi) = E_0(1 + \cos \phi) + \tfrac{1}{2}E_1(1 + \cos 3\phi) \qquad \text{eq. 1.3}$$

where E_0 is ca 2 kcal mol^{-1} and E_1 is ca 3.5 kcal mol^{-1} (Fig. 1.4).

In cases such as toluene or nitrobenzene where there is a methyl group bound to an sp^2 hybridized atom the molecular symmetry changes associated with rotation dictate a six-fold barrier which turns out to have a very low magnitude.

$$E(\phi) = \tfrac{1}{2}E_0(1 + \cos 6\phi) \qquad \text{eq. 1.4}$$

$E_0 = 0.0139$ kcal mol^{-1} for toluene.[18b]

The origins of this torsional barrier are not at all well understood. A wide variety of SCF–MO treatments of many simple systems are available[19–22] which give numerical values for barrier heights close to those found experimentally, although for each treatment the breakdown into repulsive and attractive terms varies quite considerably. Because of this, and because the

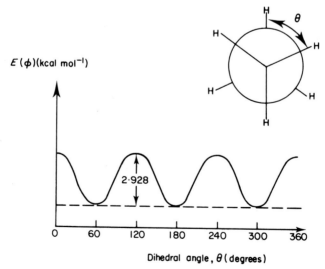

Fig. 1.3   Barrier to rotation in ethane.

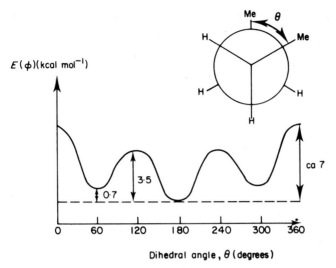

Fig. 1.4   Barrier to rotation in *n*-butane.

average organic chemist has a fear of, or lack of understanding of these quantum mechanical approaches, it is difficult to give a simple qualitative explanation of the origin of the barrier. Perhaps the most readily understood treatment is that of Allen,[19] who splits the energy terms up into one-electron and two-electron components and shows the barrier origin to result from the interactions of lone pairs, bonds and nuclei. Thus the energy difference between two conformations is given by:

$$\Delta E = \Delta T + \Delta V_{ne} + \Delta V_{ee} + \Delta V_{nn} \qquad \text{eq. 1.5}$$

where E is the total energy, T is the kinetic energy of the electrons and the terms V are the attractions of nuclei and electrons (ne) the interelectronic term (ee) and internuclear repulsion (nn).

Allen points out that the one-electron term ($V_{ne} + V_{nn} + T$) should be out of phase with the two-electron term V_{ee} as shown in Fig. 1.5. The sum of these two terms leads to the experimentally determined result. When ethane is staggered the electron–electron interactions are at a minimum. When ethane is eclipsed the electron–electron repulsions increase to a maximum. The opposite change is observed in the one-electron term ($V_{ne} + T + V_{ne}$), but the energy changes associated with this term are somewhat less. Therefore the staggered form is more stable than the *cis*.

Allen's ideas have been expanded by Eilers and Lieberles.[19c] When the components in eq. 1.5 are scaled to conform to the virial theorem $\Delta E = -\Delta T$ we have:

$$\Delta E = \tfrac{1}{2}(\Delta V_{ne} + \Delta V_{ee} + \Delta V_{nn}) \qquad \text{eq. 1.6}$$

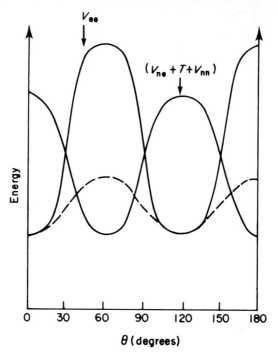

Fig. 1.5   Separation of the barrier to rotation in ethane into one-electron $(V_{ne} + T + V_{nn})$ and two-electron (V_{ee}) components. The experimental barrier is shown as a dashed line.

and the energy change can be expressed as a sum of three potential-energy components.

Their calculations are typified by the chair to twist change in cyclohexane. The values V_{ee} and V_{nn} are found to increase, whilst V_{ne} becomes more negative. The sum $(\Delta V_{ee} + \Delta V_{nn})$ is larger than the magnitude of ΔV_{ne}, giving a result in agreement with experiment and "classical" conformation theory. The changes in the terms $(\Delta V_{ne} = -1661.8, \Delta V_{ee} = 836.2, \Delta V_{nn} = 845.4$ kcal $mol^{-1})$ are enormous compared with calculated energy differences from this approach. Thus in the change from chair to twist–boat one should not attribute the energy difference solely to repulsive interactions $(\Delta V_{ee} + \Delta V_{nn})$. The change observed, ca 6 kcal mol^{-1}, is to a first approximation only about one two-hundredth of the change in repulsive interactions. It is the large compensating change in the attractive term ΔV_{ne} that keeps the total difference small.

As far as practical applications of rotation potential functions in molecular mechanics for conformational analysis are concerned, several points need to be made clear. For cycloalkanes it has been found best[23] to use the barrier to rotation in propane[24] (ca 3.4 kcal mol^{-1}) rather than the smaller barrier in ethane. The calculated geometries resulting from using this potential function are little changed but the calculated energies fit experimental results much better.[23] It would therefore seem that the rotation potential function in, for example, dimethyl ether, should be more realistic for calculations on heterocycles than that in methanol.

Non-bonded interactions

Owing to the Pauli exclusion principle, the close approach of atoms or groups that are not directly bonded will lead to repulsive forces between their electrons. Several empirical equations are in use to estimate the energy of this interaction. The weak attractive forces at large internuclear distances are generally represented by a 6th power term and the repulsive forces at short internuclear distances by either a 12th power term or an exponential term. Hendrickson[25] chose the latter course and used an equation:

$$E(r) = Ae^{-Br} - Cr^{-6} \qquad \text{eq. 1.7}$$

where $E(r)$ is the interaction energy, r is the distance separating the interacting atoms and A, B, and C are constants for the atom pair concerned. For hydrogen–hydrogen interactions Hendrickson suggests the functions:

$$E = 1.00 \times 10^4 \, e^{-4.60r} - 49.2r^{-6} \qquad \text{eq. 1.7(i)}$$

where E is in kcal mol^{-1} and r is in Å.

This function is strongly repulsive at short distances (< 200 pm) and has a feebly attractive potential energy minimum at ca 250 pm.

The Lennard–Jones potential is very often used:

$$E = \varepsilon \frac{6}{n-6} \left(\frac{r_m}{r} \right)^n - \frac{n}{n-6} \left(\frac{r_m}{r} \right)^6 \qquad \text{eq. 1.8}$$

where ε is the depth of potential energy minimum and n is usually taken as 12. Lifson[26] paramaterizes this equation for H–H interactions as

$$E = 4.5 \times 10^{-3} \left(\frac{2.936}{r} \right)^{12} - 9.0 \times 10^{-3} \left(\frac{2.936}{r} \right)^6 \qquad \text{eq. 1.8(i)}$$

where E and r are in kcal mol^{-1} and Å respectively.

The distance at which repulsive and attractive forces between two atoms are equal, is known as the van der Waals distance, which can be expressed as a sum of van der Waals radii for the atoms concerned. It should be noted, however, that the van der Waals radius of an atom is affected by the other atoms to which it is bonded and the nature of the bonding. Thus amine nitrogen will have a different radius from amide nitrogen, and carbonyl oxygen will be different from ether oxygen. The Lennard–Jones parameters will therefore vary with the nature of the system being studied.

For the conformational analysis of heterocyclic compounds it is important to note that no *satisfactory* potentials currently exist for the interactions of hydrogen atoms with the more common heteroatoms—oxygen, nitrogen, and sulphur. One of the future challenges to the subject is the setting up of such expressions. At present all that we can say is that it is easier to push a non-polar group into oxygen than into nitrogen, and that nitrogen is correspondingly softer than a methylene group.

Bond stretching

Deformations of conformation due to bond stretching are generally very small due to the large force constant involved. Lifson gives the following expression[23]

$$E(l) = 0.300(l - 153.3)^2 \text{ kcal mol}^{-1} \qquad \text{eq. 1.9}$$

whilst more recently White[17] has suggested:

$$E(l) = 0.3168(l - 152.0)^2 \text{ kcal mol}^{-1} \qquad \text{eq. 1.9b}$$

where l is bond length in pm, for bond length deformation of a C–C bond. Similar expressions are to be expected for bonds involving heteroatoms. On the basis of eq. 1.9 a stretching of 1 pm would introduce ca 0.3 kcal mol^{-1} strain. This figure corresponds to a 0.7% change in bond length. It may be calculated (see below) that a 0.3 kcal mol^{-1} strain would change a C—C—C bond angle by ca 4%, thus bond length changes, due to conformational strains, are proportionately smaller than other molecular deformations.

Bond angle deformation

In constructing a potential for the deformation of bond angles, the harmonic function of eq. 1.10 is generally employed,[23,25]

$$E(\theta) = K_\theta(\theta - \tau)^2 \qquad \text{eq. 1.10}$$

where K_θ is the force constant, θ the new angle, and τ the original undeformed angle, corresponding to minimum potential energy. It is commonly accepted for C–C–C angles that τ is not taken as tetrahedral (109.5°) but rather as the angle in acyclic hydrocarbons (ca 112.7°), as this tends to give better agreement between calculation and experiment.[23]

Lifson[23] modifies eq. 1.10 to:

$$E(\theta) = K_\theta \left(\frac{\pi}{180}\right)^2 (\theta - 67.2°)^2 \qquad \text{eq. 1.11}$$

where θ is the supplementary bond angle, to give angles in degrees, not radians, and allow for a non-tetrahedral initial geometry. Typical values of force constants are in the range 0.3–0.9 m dyn Å rad^{-2} or 20–65 kcal rad^{-2}. Westheimer[27] suggests values of 23.0, 39.6, and 57.5 kcal rad^{-2} for H—C—H, H—C—C, and C—C—C bond angles respectively. These values were employed by Hendrickson.[25] White[17] gives a more extensive set of parameters for bond angle deformation that covers both alkanes and alkenes.

It is reasonably cheap in energy to deform bond angles, and proportionately considerably cheaper than stretching bonds. The C—C—C angle in an alkane can be deformed 5° from its equilibrium value for an expenditure of only ca 0.4 kcal mol^{-1} and a 10° bending only introduces ca 1.7 kcal mol^{-1} of strain. Thus much of the conformational strain in a molecule will be accommodated, if possible, in the form of deformed bond angles.

Electrostatic interactions

In conformational analysis we are concerned with two main types of polar grouping: charged atoms or groups as in an ammonium ion, and polar bonds as in C—halogen. These different charged entities give rise to three main types of electrostatic interaction: charge–charge, charge–dipole, and dipole–dipole. Charge–charge interactions are of great importance in determining the conformations of, for example, peptide chains.

The energy of interaction of two charges e_1 and e_2 separated in a medium of dielectric constant D by a distance r is given by:

$$E = \frac{e_1 e_2}{Dr} \qquad \text{eq. 1.12}$$

If e_2 is replaced by a dipole moment μ which makes an acute angle θ with the line joining the charge to the centre of the dipole the expression then becomes

$$E = \frac{e_1 \mu \cos \theta}{Dr^2} \qquad \text{eq. 1.13}$$

This expression presupposes that r is large compared with the length of the dipole.

The potential energy of two point dipoles, μ_1 and μ_2, separated by a distance r, making angles α_1 and α_2 with their common axis and with a dihedral angle χ along their common axis (Fig. 1.6) is given by:[28]

$$E = \frac{\mu_1\mu_2}{r^3}(\cos\chi - 3\cos\alpha_1\cos\alpha_2) \qquad \text{eq. 1.14}$$

This equation was used by Smyth, Dornte, and Wilson in 1931,[29] where it seems they wrongly defined α_2 in the figure corresponding to Fig. 1.6. This mistake was pointed out by Lehn and Ourisson,[30] who also extended eq. 1.14 to the case of two dipolar atoms e.g. halogen or carbonyl oxygen adjacent on a common C–C bond. Typically, energies of electrostatic interactions in molecules vary from 0 to 2 kcal mol^{-1} both when calculated from eq. 1.14

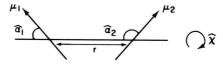

Fig. 1.6   Geometric relationship between two point dipoles for eq. 1.14.

and when measured experimentally. However, great difficulties are encountered in calculating all electrostatic interactions inside molecules. First, the above expressions are only valid for point dipoles. On a molecular scale the electron distribution giving rise to, say, a carbon–halogen bond dipole cannot be considered as a point dipole. This is the more so since it may be interacting with another dipole at a distance comparable with the carbon–halogen bond length. Secondly, if one accepts the point dipole approximation and assigns the dipole a position in the polar bond, one cannot justify any chosen point of action as superior to another, hence one does not know where to place the dipole. Thirdly, it is not clear what dielectric constant should be employed. The expression in eq. 1.14 relates to free space but maybe a more hydrocarbon-like dielectric constant would be appropriate. Thus eqs 1.12 to 1.14 can only be used as qualitative indicators of effects. Quantitatively they are not much use.

There will of course be interactions involving molecules that have quadrupoles and octupoles, etc. However, as we have seen the higher the order of the pole involved the more rapidly the interactions fall off with distance. Thus charge–charge interactions vary as r^{-1}, charge–dipole as r^{-2}, dipole–dipole as r^{-3}. Similarly dipole–quadrupole interactions vary as r^{-4} and fall off very rapidly with distance. Quadrupole and octupole interactions therefore are of little importance for the ensuing discussion.

Acyclic Compounds

As described earlier the barrier to rotation in ethane is found to be sinusoidal with a height of 2.928 kcal mol^{-1}. In the isoelectronic first-row compounds, ethane to hydrogen peroxide, a striking variation in shape and magnitude of barriers is found. Firstly, ethane, methylamine, and methanol show three-fold barriers (eq. 1.2) with decreasing barrier heights. Secondly, the compounds with two heteroatoms, hydrazine, hydroxylamine and hydrogen peroxide, either show, or are calculated to have barriers up to an

TABLE 1.1   Selected barriers to internal rotation

Formula	Compound	Barrier height (kcal mol^{-1})	Ref.
CH_3—CH_3	Ethane	2.928	18a
CH_3—CH_2F	Ethylfluoride	3.33	31
CH_3—CH_2Cl	Ethylchloride	3.68	32
CH_3—CH_2Br	Ethylbromide	3.68	32
CH_3—CH_2I	Ethyliodide	3.22	33
CH_3—CH_2CH_3	Propane	3.575	34
CH_3—$CH(CH_3)_2$	2-Methylpropane	3.9	35
CH_3—$C(CH_3)_3$	2,2-Dimethylpropane	4.7	36
CH_3—CH_2OH	Ethanol	3.3	37
CH_3—CH_2SH	Ethanethiol	3.31	38
CH_3—CH_2CHO	Propionaldehyde (for planar *cis* form)	2.28	39
CH_3—CHO	Acetaldehyde	1.18	40
CH_3—$COCH_3$	Acetone	0.78	32, 41
CH_3—$CH=CH_2$	Propene	1.99	42
CH_3—NH_2	Methylamine	1.98	43
CH_3—$NHCH_3$	Dimethylamine	3.62	32, 44
CH_3—$N(CH_3)_2$	Trimethylamine	4.41	45
CH_3—$N=CH_2$	*N*-Methylmethyleneimine	1.97	46
CH_3—PH_2	Methylphosphine	1.96	47
CH_3—$P(CH_3)_2$	Trimethylphosphine	2.6	48
CH_3—OH	Methanol	1.11	49
CH_3—OCH_3	Dimethylether	2.7	32, 44, 50
CH_3—SH	Methylthiol	1.27	51
CH_3—SCH_3	Dimethylsulphide	2.53	52
CH_3—$SOCH_3$	Dimethylsulphoxide	2.94	53
CH_3—SeH	Methylselenol	1.01	54
CH_3—$SeCH_3$	Dimethylselenide	1.50	55
NH_2—NH_2	Hydrazine	11.88/3.70^a	19
NH_2—OH	Hydroxylamine	11.95^a	19
HO—OH	Hydrogen peroxide	7.0/1.1	56

^{a} Values from MO calculations, not from experiment.

order of magnitude greater, with a two-fold function. These results are shown in Table 1.1. These striking differences are clearly going to be of importance when comparing the effects of twisting carbon–carbon, carbon–heteroatom and heteroatom–heteroatom bonds. This point will be discussed in Chapter 2.

For substituted derivatives of ethane, methylamine and methanol, the barrier remains three-fold, but different maxima and minima may occur as in the function of n-butane. In these compounds the barriers are generally greater than in the parents and they are better models for cyclic compounds. With heavily substituted bonds the barriers can be quite high. For example, in 2-chloro-2,3-dimethylbutane (11)[57] the largest barrier of 8.2 kcal mol^{-1} for (11a) → (11b) is probably strongly influenced by the repulsive non-bonded interactions between two pairs of methyl groups in the transition state. Non-bonded interactions also affect the preferred conformation here. Since a methyl–chlorine *gauche* interaction is ca 0.0–0.6 kcal mol^{-1} whereas the methyl–methyl *gauche* interaction is larger (0.7 kcal mol^{-1}), the conformation (11c) with the minimum number of methyl–methyl interactions and the maximum number of methyl–chlorine interactions is preferred ($\Delta G = 0.3$ kcal mol^{-1}). The approximate rotation potential function for this molecule is shown in Fig. 1.7.

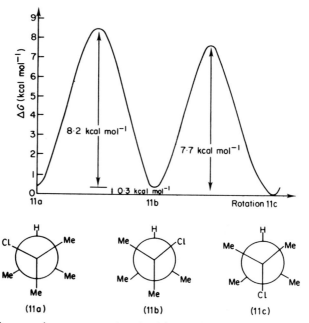

Fig. 1.7   Energy changes associated with rotation about the central bond in 2-chloro-2,3-dimethylbutane (Ref. 57).

Conformational preferences about bonds between sp^3 and sp^2 hybridized atoms show some interesting trends. In most compounds the *cis* arrangement (12) is preferred. For example, propionaldehyde (13) and methylethylketone (14) prefer conformations in which the CH$_3$ of the ethyl group and the carbonyl oxygen are *cis* about the central C–C bond. In these cases the less stable conformations have *gauche* (120° dihedral angle) arrangements of methyl and carbonyl. 1-Butene (15) prefers the *gauche* conformation of double bond and methyl. An exception to the general rule is fluoroacetone which prefers a *trans* (180° dihedral angle) arrangement of fluorine and carbonyl (16) probably because of severe dipolar interactions in other conformations.

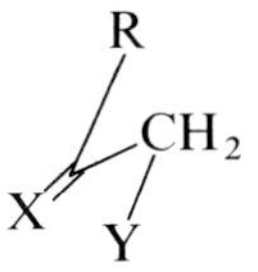

(12) 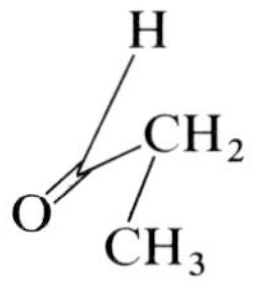   (13) 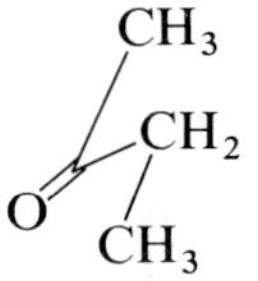   (14)

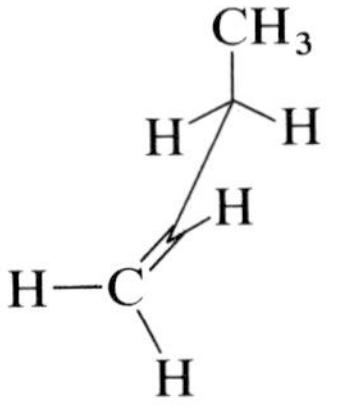

(15) 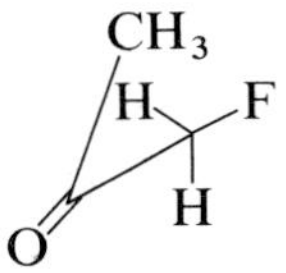   (16)

Alicyclic Compounds

The conformational analysis of alicyclic compounds, especially that of cyclohexane which is the case we shall discuss, has been extensively studied and is generally well understood in terms of the effects discussed earlier.

Cyclohexane derivatives generally have an overwhelming preference for chair conformations; however, the existence of non-chair conformations of six-membered rings is now well documented.[58] For any six-membered ring there are certain common geometrical considerations. In the case of cyclohexane the following three conditions are approximately true: (1) the atoms are joined in a ring; (2) all bond lengths are equal; (3) all C—C—C bond angles are equal. It is possible to demonstrate mathematically that two sets of arrangements fulfil these conditions. The first has only one member and is rigid, being the well-known chair conformation of symmetry D_3d. The second set has an infinite number of members which are continuously inter-convertible, all of which possess a C_2 axis, and includes the twist form (D_2) and boat form (C_{2v}). This set is known as the boat–twist family, and their interconversion mode is called pseudorotation.[58]

For cyclohexane, in reality, other considerations apply. Bond lengths and bond angles vary with molecular vibrations, and energy is also required for the rotation about C—C bonds that occurs during pseudorotation. Since one of the three conditions set out above must be broken to pass from a chair to the boat–twist family during the ring inversion process, some bond length and bond angle changes occur during ring inversion. Moreover, non-bonded interactions and torsional forces make the twist about 1 kcal mol^{-1} more stable than the boat. The twist is about 5 kcal mol^{-1} higher in energy than the chair.[58]

When heterocycles are considered, condition (2), that of equal bond lengths, is broken. It is possible to show that in these compounds, unless the molecule has an axis of symmetry, the previously infinite boat–twist family now has only certain allowable positions. This does not mean that pseudo-rotation inside the boat–twist family cannot occur in heterocycles, for here again small variations in bond lengths and angles, at a modest expense in energy, are permissible in real molecules.

Two types of positions are available for substituents in cyclohexane. These are the well-known axial and equatorial locations. Axially substituted conformations are generally less stable than equatorial conformations because of non-bonded repulsive interactions across the top of the molecule. Some energy differences are given in Table 1.2.

In general the elements of the first two rows in the Periodic Table, with relatively short bonds to carbon and low polarizability of their electrons, show larger axial equatorial free energy differences than the heavier elements which have longer bonds to carbon and higher polarizabilities.

The effect of van der Waals radius (i.e. effective size of the atom) is seen in a comparison of the values for oxygen (0.5–0.7 kcal mol^{-1}) with those of sulphur (1.1–1.2 kcal mol^{-1}). The radius of sulphur (185 pm) is greater than the radius of oxygen (140 pm)* by an amount greater than the difference in their bond lengths (181 pm − 142 pm = 39 pm).

The effect of bond length and polarizability is shown in the value for trichlorosilyl (0.61 kcal mol^{-1}) compared with t-butyl (>4 kcal mol^{-1}). Although both groups are approximately the same size the carbon–silicon bond (188 pm) is considerably longer than the C—C bond (153 pm) moving

---

* Values from Pauling. (1960). "The Nature of the Chemical Bond," Cornell University Press, Ithaca.

TABLE 1.2   Selected conformational energy differences in cyclohexane derivatives (kcal mol^{-1})

Group	ΔG	T	ΔH	ΔS	Ref.
CH_3	1.56–1.75	25–36	1.60	−0.25	59–62
CH_2CH_3	1.67–1.86	25–35	1.65	−0.7	59, 61
$CH(CH_3)_2$	2.30–2.62	25–35	2.4–2.5	−0.1–+0.4	59, 61
F	0.28	−80	—	—	63–65
Cl	0.53	−80	—	—	63, 64, 66
Br	0.48	−80	—	—	63, 66, 67
I	0.47	−80	—	—	63, 66
OCH_3	0.6	?	—	—	68
OCD_3	0.55	−82	—	—	63, 69
OH	0.6–0.9	80	—	—	70–73
SH	1.2	−80	—	—	63
SCD_3	1.07	−79	—	—	63
HgOAc	0.0	−79	—	—	63
$SiCl_3$	0.61	−80	—	—	63

the interacting centres further apart. There may also be some contribution from the greater polarizability of the electrons on chlorine and silicon.

The fifth-row element mercury has been found to have a zero energy difference. This presumably arises from the long carbon–mercury bond (207 pm) and the high polarizability of its atoms.

There are many review articles relevant to the subject matter in this chapter. The reader may care to consult references 74–78.

References

1a.   J. H. van't Hoff. (1875). *Bull. Soc. Chim. France* **23**, 295. (The original version appeared in Dutch in 1874.)

1b.   J. A. Le Bel. (1874). *Bull. Soc. Chim. France* **22**, 337.

2.   F. G. Riddell and M. J. T. Robinson. (1974). *Tetrahedron* **30**, 2001.

3.   H. Sachse. (1890). *Chem. Ber.* **23**, 1363; (1892). *Z. Phys. Chem.* **10**, 203.

4.   E. Mohr. (1918). *J. Prakt. Chem.* **98**, 315.

5.   W. Hückel. (1925). *Justus Liebigs Ann. Chem.* **441**, 42.

6.   See for example: J. Boeseken and H. G. Derx. (1921). *Rec. Trav. Chim.* **40**, 519, 529; and also J. Boeseken and W. J. F. de Rijck van der Gracht. (1937). *Rec. Trav. Chim.* **56**, 1203.

7.   W. N. Haworth. (1929). "The Constitution of Sugars", 90. Edward Arnold, London.

8.   R. G. Dickinson and C. Bilicke. (1928). *J. Am. Chem. Soc.* **50**, 764.

9.   K. W. F. Kohlrausch, A. W. Reitz and W. Stockmair. (1936). *Z. Phys. Chem.* **B32**, 229.

10. O. Hassel. (1943). *Tidsskrift for Kjemi og Bergvaesen og Metallurgi* 3, 32. See also *Quart. Rev.* (1953) 7, 221.
11. D. H. R. Barton. (1950). *Experientia* 6, 316.
12. For a review of their contributions to Conformational Analysis see E. L. Eliel. (1969). *Science (N.Y.)* 166, 718.
13. For further background the reader should consult:
    a. E. L. Eliel. (1962). "Stereochemistry of Carbon Compounds", McGraw-Hill, New York.
    b. M. Hanack. (1965). "Conformation Theory", Academic Press, New York and London.
    c. E. L. Eliel, N. L. Allinger, S. J. Angyal and G. A. Morrison. (1965). "Conformational Analysis", Interscience, New York.
14. D. H. R. Barton and R. C. Cookson. (1956). *Quart. Rev.* 10, 44.
15. F. G. Riddell. (1967). *Quart. Rev.* 21, 364.
16. IUPAC Rules for the Nomenclature of Organic Chemistry. Section E, Fundamental Stereochemistry. (1976). *Pure Appl. Chem.* 45, 13.
17. D. N. J. White and M. J. Bovill. (1977). *J. C. S. Perk. II.* 1610.
18a. S. Weiss and G. E. Le Roi. (1958). *J. Chem. Phys.* 29, 340.
18b. H. R. Rudolf, H. Dreizler, A. Jaeschke and P. Windling. (1967). *Z. Naturforsch* 22a, 940.
19a. W. H. Fink and L. C. Allen. (1967). *J. Chem. Phys.* 46, 2261, 2276.
19b. W. H. Fink, D. C. Pan and L. C. Allen. (1967). *J. Chem. Phys.* 47, 895.
19c. J. E. Eilers and A. Lieberles. (1975). *J. Am. Chem. Soc.* 97, 4183.
20. L. Pedersen and K. Morokuma. (1967). *J. Chem. Phys.* 46, 3941.
21. R. M. Pitzer and W. N. Lipscomb. (1963). *J. Chem. Phys.* 39, 1995.
22. R. M. Pitzer. (1964). *J. Chem. Phys.* 41, 2216.
23. M. Bixon and S. Lifson. (1967). *Tetrahedron* 23, 769.
24. G. B. Kistiakowsky and W. W. Rice. (1940). *J. Chem. Phys.* 8, 610.
25. J. B. Hendrickson. (1961). *J. Am. Chem. Soc.* 83, 4537.
26. S. Lifson and A. Warshel. (1968). *J. Chem. Phys.* 49, 5116.
27. F. H. Westheimer. (1956). In "Steric Effects in Organic Chemistry", (Ed. M. S. Neuman), Chap. 12. Wiley, New York.
28. J. H. Jeans. (1925). "The Mathematical Theory of Electricity and Magnetism", 5th ed, eq. 354 on p. 379. Cambridge University Press, London.
29. C. P. Smyth, R. W. Dornte and E. B. Wilson, Jr. (1931). *J. Am. Chem. Soc.* 53, 4242.
30. J. M. Lehn and G. Ourisson. (1963). *Bull. Soc. Chim. France* 1113.
31. G. Sage and W. Klemperer. (1963). *J. Chem. Phys.* 39, 371.
32. K. D. Möller, A. R. De Meo, D. R. Smith, and L. H. London, (1967). *J. Chem. Phys.* 47, 2609.
33. T. Kasuya. (1960). *J. Phys. Soc. Jpn* 15, 1273.
34. K. S. Pitzer. (1944). *J. Chem. Phys.* 12, 310.
35. D. R. Lide, Jr. and D. E. Mann. (1958). *J. Chem. Phys,* 29, 914.
36. K. S. Pitzer. (1937). *J. Chem. Phys.* 5, 473.
37. G. M. Barrow. (1952). *J. Chem. Phys.* 20, 1739.
38. J. P. McCullough, D. W. Scott, H. L. Finke, W. N. Hubbard, M. E. Gross, C. Katz, R. E. Pennington, J. F. Messerly and G. Waddington. (1952). *J. Am. Chem. Soc.* 74, 280.
39. S. S. Butcher and E. B. Wilson, Jr. (1964). *J. Chem. Phys.* 40, 1671.
40. W. G. Fately and F. A. Miller. (1961). *Spectrochim. Acta* 17, 875.

41. R. Nelson and L. Pierce. (1965). *J. Molec. Spectrosc.* **18**, 344.
42. E. Hirota. (1966). *J. Chem. Phys.* **45**, 1984.
43. D. R. Lide, Jr. (1957). *J. Chem. Phys.* **27**, 343.
44. J. P. Perchard, M. T. Ford and M. L. Josien. (1964). *J. Chim. Phys.* **61**, 632.
45. D. R. Lide, Jr. and D. E. Mann. (1958). *J. Chem. Phys.* **28**, 572.
46. J. T. Yardley, J. Hinze and R. F. Curl, Jr. (1964). *J. Chem. Phys.* **41**, 2562.
47. T. Kojima, E. L. Breig and C. C. Lin. (1961). *J. Chem. Phys.* **35**, 2139.
48. L. S. Bartell and L. O. Brockway. (1960). *J. Chem. Phys.* **32**, 512.
49. J. D. Swalen. (1955). *J. Chem. Phys.* **23**, 1739.
50. U. Blukis, P. H. Kasai and R. J. Myers. (1963). *J. Chem. Phys.* **38**, 2753.
51. T. Kojima and T. Nishikawa. (1957). *J. Phys. Soc. Jpn* **12**, 680.
52. K. D. Moller and H. G. Andresen. *J. Chem. Phys.* (1962), **37**, 1800; (1963), **39**, 17.
53. H. Dreizler and G. Dendl. (1965). *Z. Naturforsch.* **20a**, 1431.
54. C. H. Thomas. (1973). *J. Chem. Phys.* **59**, 70.
55. J. F. Beecher. (1966). *J. Molec. Spectrosc.* **21**, 414.
56. R. H. Hunt, R. A. Leacock, C. W. Peters and K. T. Hecht. (1965). *J. Chem. Phys.* **42**, 1931.
57. J. E. Anderson and H. Pearson. (1973). *J. C. S. Perk II* 960.
58. G. M. Kellie and F. G. Riddell. (1974). "Topics in Stereochemistry", Vol. 8. 225. (Eds Eliel and Allinger), Wiley Interscience, New York.
59. B. J. Armitage, G. W. Kenner and M. J. T. Robinson. (1964). *Tetrahedron* **20**, 747.
60. F. A. L. Anet, C. H. Bradley and G. W. Buchanan. (1971). *J. Am. Chem. Soc.* **93**, 258.
61. E. L. Eliel and T. J. Brett. (1965). *J. Am. Chem. Soc.* **87**, 5039.
62. J. C. Celotti, J. Reisse and G. Chiurdoglu. (1966). *Tetrahedron* **22**, 2249.
63. F. R. Jensen, C. H. Bushweller and B. H. Beck. (1969). *J. Am. Chem. Soc.* **91**, 344.
64. A. J. Berlin and F. R. Jensen. (1960). *Chem. Ind.* 998.
65. F. A. Bovey, E. W. Anderson, F. P. Hood and R. L. Kornegay. (1964). *J. Chem. Phys.* **40**, 3099.
66. W. C. Neikam and B. P. Dailey. (1965). *J. Chem. Phys.* **38**, 445.
67. F. R. Jensen and L. H. Gale. (1960). *J. Org. Chem.* **25**, 2075.
68. G. W. Buchanan, D. A. Ross and J. B. Stothers. (1966). *J. Am. Chem. Soc.* **88**, 4301.
69. F. R. Jensen and C. H. Bushweller. (1966). *J. Am. Chem. Soc.* **88**, 4279.
70. E. L. Eliel and S. H. Schroeter. (1965). *J. Am. Chem. Soc.* **87**, 5031.
71. E. L. Eliel, S. H. Schroeter, T. J. Brett, F. J. Biros and J. C. Richer. (1966). *J. Am. Chem. Soc.* **88**, 3327.
72. E. L. Eliel and M. N. Rerick. (1960). *J. Am. Chem. Soc.* **82**, 1367.
73. M. Tichy, F. Sipos and J. Sicher. (1966). *Coll. Czech. Chem. Comm.* **31**, 2889.
74. E. L. Eliel. (1975). *J. Chem. Educ.* **52**, 762.
75. J. P. Lowe. (1968). *Progr. Phys. Org. Chem.* **6**, 1.
76. J. A. Hirsch. (1967). *In* "Topics in Stereochemistry," Vol. 1, (Eds Eliel and Allinger), 199. Wiley Interscience, New York.
77. F. R. Jensen and C. H. Bushweller. (1971). *Adv. Alicyclic Chem.* **3**, 140.
78. J. Dale. (1976). *In* "Topics in Stereochemistry," Vol. 9, (Eds Eliel and Allinger), 199. Wiley Interscience, New York.

2

Special Conformational
Considerations for Heterocycles

In Chapter 1 we pointed out that a complete molecular orbital study of a molecule should give us the detailed conformational information we require. In practice, however, it is currently difficult for the non-expert to decide which of the approximate MO approaches should be adopted, and the full MO calculation not involving approximations is too complex to be handled economically by current technology. In fact, a classical approach is generally adopted which divides molecular forces into components for which the average chemist has an intuitive grasp. Bond stretching and bond angle bending, for example, are readily appreciated concepts that can be transposed from the macroscopic world in which we live to the microscopic world of the molecules we study. Chapter 1 discussed these conformational directing forces in the context of some well-known examples from acyclic and alicyclic conformational analysis. In this chapter we shall extend these concepts into the heterocyclic series.

Bond Lengths, van der Waals Radii and Non-Bonded Interactions

The van der Waals radius of an atom or group is an ill-defined quantity. Van der Waals radii can be derived in a number of different ways and the

TABLE 2.1   Van der Waals Radii (after Pauling) (pm)

		H	120		
N	150	O	140	F	135
P	190	S	185	Cl	180
As	200	Se	200	Br	195
Sb	220	Te	220	I	215

Radius of a methyl group, 200.
Half thickness of an aromatic molecule, 170.

results are not altogether consistent. In an attempt to maintain internal consistency in this book we shall use the van der Waals radii derived by Pauling which are presented in Table 2.1.

Table 2.1 shows several interesting features. For the atoms and groups present as the most common components of reduced heterocyclic systems, methylene, nitrogen and oxygen, the radii are ca 200, 150, and 140 pm respectively.* Since the van der Waals radius is an attempt at estimating how close one group can approach another it is clear that a given non-polar probe can get closer to oxygen than to nitrogen or methylene, and closer to a nitrogen atom than to a methylene group. Alternatively stated, the magnitude of the repulsion between a probe and these atoms is oxygen < nitrogen < methylene for a given distance.

We can alternatively visualize this in terms of the repulsive potentials presented in Chapter 1 (eqs 1.7 and 1.8).

Take a Lennard–Jones function:

$$E(r) = \frac{A}{r^{12}} - \frac{B}{r^6} \qquad \text{eq. 2.1}$$

when we are operating in the repulsive region of the potential A/r^{12} is the dominant term and B/r^6 has a negligible effect. The van der Waals radii therefore tell us that the order in which the A terms fall is oxygen < nitrogen < methylene. The repulsive potentials will then be roughly as outlined in Fig. 2.1. Currently, satisfactory potential functions are not available for application in heterocyclic chemistry, and one of the next main goals of force field calculations should be to develop suitable potential functions.

The distances between groups that have a non-bonded interaction are determined by the molecular geometry i.e. by bond lengths, bond angles, and torsion angles. Of these, probably the most important for cyclic compounds is the bond length. It is worth noting that the shortest bond to carbon

* This treats methylene as having the same radius as methyl. In practice this may not be exactly correct but in our qualitative argument the result is the same.

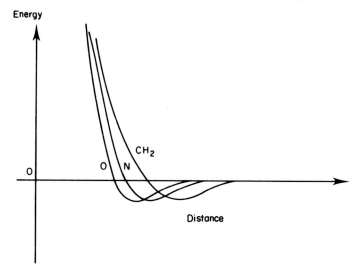

Fig. 2.1   Relative van der Waals potential functions as a non-polar probe approaches ether oxygen, amine nitrogen or methylene groups.

found in a reduced heterocycle is C—O (ca 142 pm), the next C—N (ca 148 pm) compared with C—C (ca 153 pm). Longer bonds to heteroatoms are C—S (ca 181 pm) and C—P (ca 185 pm). Bonds between heteroatoms have not been as extensively investigated, but typical values are O—O (ca 148 pm), N—O (ca 146 pm), and N—N (ca 145 pm). From these values we may deduce that changing a C—C bond between two repulsively inter-acting centres to C—O or to C—N should increase the repulsive interaction, but changing it to C—S or C—P should decrease it, provided that no other factors intervene.

We can now make some qualitative predictions about the effect of introduc-ing a heteroatom into a six-membered ring. If we imagine a cyclohexane ring with an axial methyl group we can go around the ring replacing each methyl-ene group by a heteroatom and deduce qualitatively what the effects might be.

For the first case (Fig. 2.2) if X is oxygen or nitrogen the distance between the axial methyl and H_1 has been decreased, thus enhancing the repulsion between these groups and increasing the axial equatorial free energy differ-ence. The effect will be more marked than expected for oxygen because the

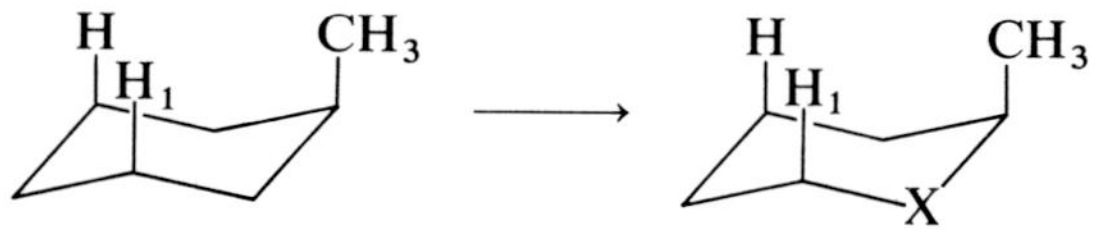

Fig. 2.2   Change to axial methylcyclohexane on introducing an α heteroatom.

C—O—C bond angle (110°) is slightly less than the normal C—C—C bond angle (113°) moving the methyl and hydrogen even closer together. By contrast if X is sulphur the distance is expected to increase because of the longer C—S bond lengths, although in practice the narrowing of the C—S—C angle to 100° offsets this and the interactions are comparable to those in cyclohexane.

For the next case (Fig. 2.3) two changes are involved: the removal of the axial C—H bond which is replaced by a pair of electrons, and an alteration in the methyl—X distance due to the bond length change. For oxygen the distance decrease is substantially less than the decrease in van der Waals radius. This reduces the axial equatorial free energy difference. For nitrogen the changes in distance and van der Waals radii are somewhat smaller, so a lesser effect is expected. For sulphur the increase in the bond length and the slightly lower radius of sulphur join in reducing the conformational energy difference of the methyl group.

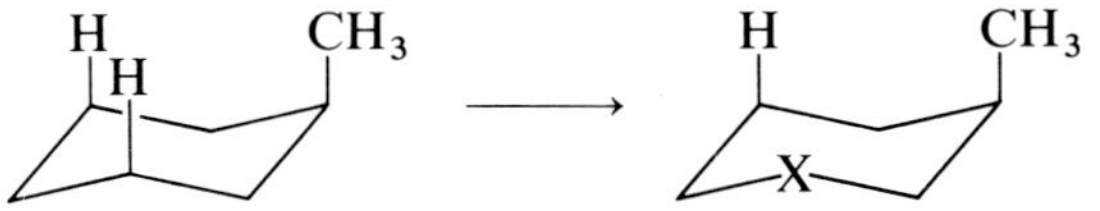

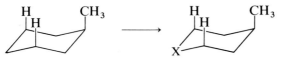

Fig. 2.3   Change to axial methylcyclohexane on introducing a β heteroatom.

When the heteroatom is moved further round the ring minimal effects are expected (Fig. 2.4). The smaller atoms O and N may "pinch in" the far end of the molecule from the methyl, raising the repulsions slightly. The larger atoms may have the opposite effect which in the case of sulphur may be counterbalanced by the smaller C—S—C angle.

The above analysis of repulsive interactions is by and large substantiated by experimental results discussed in Chapters 4–6.

The above discussion has always assumed that the interacting groups are in the repulsive region of the van der Waals potential function. It is possible for slightly larger distances to occur in molecules that place the groups in the feebly attractive energy minimum of the curve. This may account for the conformational preferences of several thian derivatives for which axial 5-substituents are preferred.[2]

A further effect of differences in bond lengths on repulsive interactions is seen in chair–twist energy differences. 1,3-Dioxan is a smaller molecule

Fig. 2.4   Change to axial methylcyclohexane on introducing a γ heteroatom.

ΔG_{ct} ca 8.5   ca 5.5   ca 3.5 kcal mol^{-1}

1,3-Dioxan   Cyclohexane   1,3-Dithian

Fig. 2.5   Chair–twist free energy differences in 1,3-dioxan, cyclohexane and 1,3-dithian.

than cyclohexane, which in turn is smaller than 1,3-dithian (Fig. 2.5). The differences in size also reflect the differences in chair–twist free energy differences. The short bond lengths in 1,3-dioxan bring about repulsive interactions in the twist that are greater than in cyclohexane. Conversely the longer bond lengths in 1,3-dithian reduce the repulsive interactions in the twist form with respect to those in cyclohexane, and the energy of the twist forms with respect to the chairs is therefore dioxan > cyclohexane > dithian.

Torsional Interactions

Differences in torsional interactions between heterocycles and alicycles arise because, as discussed in Chapter 1, the torsional potentials along heteroatom–heteroatom and carbon–heteroatom bonds differ from those about carbon–carbon bonds.

Heteroatom–heteroatom bonds

The experimental potential function for hydrogen peroxide[4] is shown in Fig. 2.6(a) and the calculated functions for hydroxylamine and hydrazine in Figs 2.6(b) and (c) respectively.[5] Experimental results on methylated hydroxylamines[6] and on methyl hydrazine[7] qualitatively confirm the shape of the potential function. It is apparent, therefore, that rotation potential functions for these compounds display two maxima and two minima on the 360° rotation itinerary and that the barrier heights are up to an order of magnitude greater than for carbon–carbon, or carbon–heteroatom single bonds. It is also seen that dihedral angles of 0, 90°, and 180° are preferred by these molecules as opposed to the *gauche* and *trans* arrangements (60° and 180°) in the more common cases.

Introduction of a heteroatom–heteroatom bond into a molecule will therefore have two possible conformational effects: first, the barrier to any

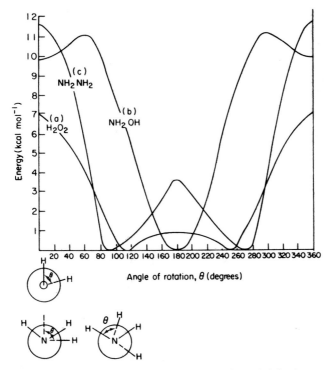

Fig. 2.6   Torsion potentials about the central bonds in (a) hydrogen peroxide, (b) hydroxylamine, (c) hydrazine (Refs 4, 5, 6, 7).

torsional process about the bond will be raised, and secondly, torsion angles about these bonds when incorporated into molecules, will be altered to accommodate the different position of minimum strain energy. The first of these effects is seen in the barrier to ring inversion in 1,2,4,5-tetraoxan derivatives (1)[8] (ca 15 kcal mol^{-1}) and in 5,5-dimethyl-1,3,2-dioxathian 2)[9] (ca 12.5 kcal mol^{-1}) compared with cyclohexane and 1,3-dioxan as models (ca 10.3 and 9.7 kcal mol^{-1} respectively). The second effect is seen n the preferred torsion angles about O—O bonds in the 1,2,4,5-tetraoxan ing (1) of ca 64°,[10] and the observed N—O torsion angle in the tetrahydro-1,2-oxazine ring (3) of 67°.[11] These values are greater than in cyclohexane

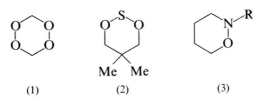

(1)   (2)   (3)

(ca 55°) because the molecules enlarge the endocyclic torsion angle in an attempt to get as close as possible to the minimum energy position for rotation about the O—O and N—O bonds. Both the enhancement of the ring inversion barrier and the alteration of ring dihedral angles from "normal" values are shown very nicely in the studies of cyclic hydrazines by Nelsen and co-workers.[12]

Carbon–heteroatom bonds

It is difficult to find an unambiguous example where the different rotation potential function about a carbon–heteroatom bond modifies the behaviour of a heterocycle with respect to the corresponding alicycle. The barriers to ring inversion in four-membered rings may provide an example. These barriers are composed largely of a mixture of angle deformation and torsional strain energies. In the planar transition state torsional interactions are at a maximum because each bond is perfectly eclipsed. As the molecule becomes non-planar to relieve the torsional interactions, angle strain builds up as the bond angles decrease from their maximum value of ca 90° (exactly 90° in the case of cyclobutane). Since the barrier to rotation in methanol is one-third that in ethane it might be expected that oxetane should show a lower barrier to ring inversion than cyclobutane. This is indeed the case (<0.1 kcal mol^{-1} vs ca 1.4 kcal mol^{-1}).[13] The barrier in azetidine (1.26 kcal mol^{-1}) is intermediate between these two, in line with the barrier to rotation in methylamine being between those of ethane and methanol.

Carbon–carbon bonds

Rotation about carbon–carbon bonds becomes important for us when heteroatoms are attached to either end. Two cases have been noted where anomalies apparently exist. When the substituents are small, highly electronegative atoms such as fluorine and oxygen, the *gauche* rather than the *trans* orientation is preferred.[14] Various explanations have been put forward for this so-called "*gauche* effect" none of which either appear satisfactory to the current author, or give results that reproduce the experimental data. However, conventional molecular mechanics calculations using carefully chosen steric interaction potentials and a real standardized non-tetrahedral geometry can give a satisfactory reproduction of observed rotation barriers and *gauche–trans* energy differences.[15] It appears, therefore, that the so-called "*gauche* effect" can be adequately explained.

On the other hand enhanced repulsions appear to exist beyond those readily accounted for by steric and polar factors when two atoms of the second (or lower) row of the periodic table are *gauche* to each other.[16-18] This effect was called the "hockey sticks" effect in the earlier papers of Zefirov[16] who first mentioned the effect, but is now referred to as the "*gauche-repulsive*" effect. This enhanced repulsion is found also between *gauche* oxygen and sulphur and has been investigated in some detail by Eliel.[18]

Electrostatic Interactions

The polar nature of bonds in heterocyclic compounds means that electrostatic interactions should be conformationally very important. This is exemplified in a variety of systems, but perhaps most elegantly in the series of piperidinium salts (4). In these compounds charge dipole interactions stabilise the chair conformation which has the axial electronegative substituent.[19] In contrast non-polar groups such as 4-methyl behave normally and prefer to be equatorial (ΔG ca 2 kcal mol^{-1}).

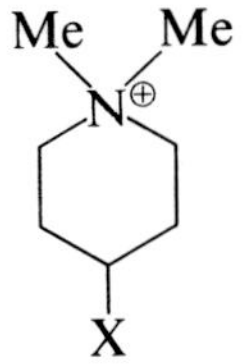

$$X = OH, Cl,$$
$$OAc \text{ or } OBz$$

(4)

The best-known supposed example of dipole–dipole interactions in heterocyclic systems is the anomeric effect. Qualitatively this effect can be explained on the basis of dipolar interactions of an electrostatic nature, although modern theory treats the effects in terms of orbital interactions.

The anomeric effect, first recognized in sugars, but now accepted as a general feature of tetrahydropyran chemistry, is the preference of a polar 2-substituent e.g. halogen or methoxy for the axial position in a tetrahydropyran ring. In the dipolar explanation it is the favourable interaction in conformation (5) where the dipoles are antiparallel, versus the negligible interactions in (6) or the unfavourable parallel interactions in (7) that results in the axial preference.

An excellent review of the anomeric effect is given in Ref. 20. In this article the authors point out that in 2-halo-1,4-dioxans, thioxans, and dithians, which show exceptionally strong anomeric effects, the C(2)–heteroatom distance (adjacent to the axial halogen) is considerably shorter than the

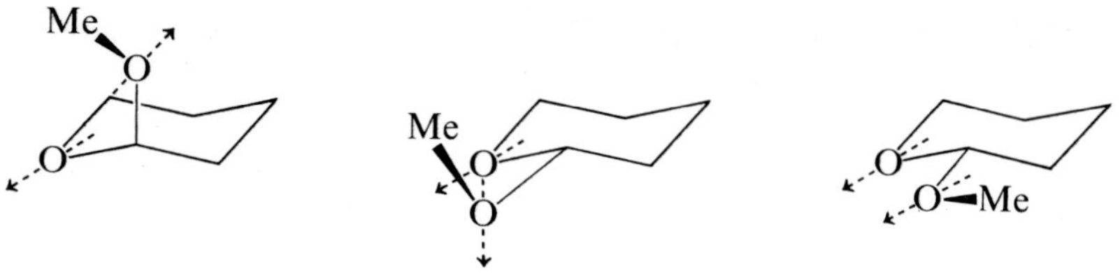

(5)   (6)   (7)

C(6)–heteroatom distance. Thus in *trans*-2,3-dichloro-1,4-dioxan the C(2)–O distance is 138 pm whereas the C(6)–O distance is normal for an ether at 143 pm. Conversely, the bond lengths to the axial halogen atoms are longer than normal. These abnormalities in bond lengths suggest a strengthening of the C(2)–O bond and a weakening of the C(2)–halogen bond. This can be accounted for by delocalization of the non-bonding σ orbital of the C(2)–halogen bond (Fig. 2.7). Molecular orbital calculations have provided support for this interpretation.[21]

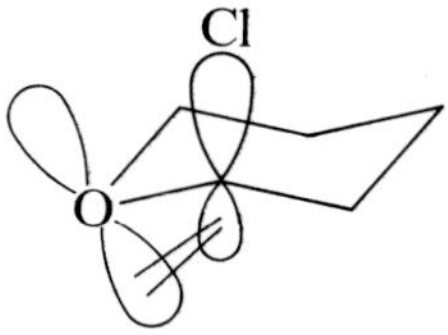

Fig. 2.7   Anomeric orbital interaction in axial 2-chlorotetrahydropyran (Ref. 20).

A reverse anomeric effect has been postulated in systems containing a strongly electron-withdrawing group e.g. $-\overset{+}{N}R_3$ attached to the C(2) position in tetrahydropyran systems.[22] A particularly striking manifestation of the reverse anomeric effect is seen in *N*-glycosylimidazoles where protonation or quaternization of the imidazole ring changes its preferred orientation from axial to equatorial (Fig. 2.8).[23] Some evidence of a reverse anomeric effect in 1,3-dioxans has been produced by Eliel.[23]

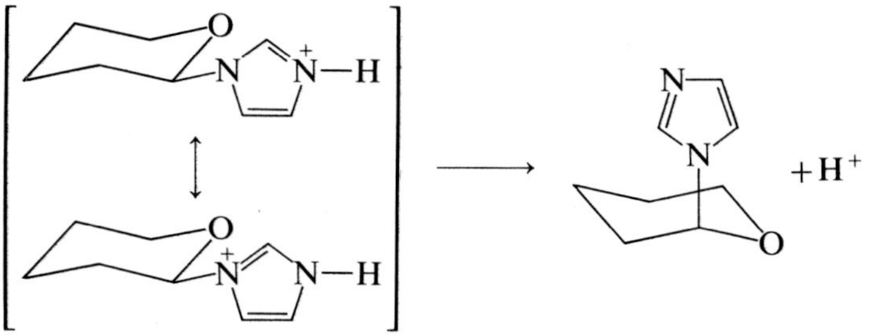

Fig. 2.8   Conformational change brought about by protonation in *N*-glycosylimidazoles (Ref. 23).

Hydrogen Bonding

A hydroxyl group in cyclohexane has an equatorial preference of about 1 kcal mol^{-1}. In heterocyclic compounds axial hydroxyl groups can in many cases hydrogen bond to the ring heteroatom(s). Since hydrogen bond strengths are often greater than 1 kcal mol^{-1} the axial conformation is frequently found to be the more stable for many heterocyclic compounds. In pyranose sugars with many hydroxyl groups, hydrogen bonding of the hydroxyl groups with each other and with the ring oxygen can be the dominating conformational influence.

A change of solvent from non-polar to hydrogen bonding (e.g. H_2O or CH_3OH) can influence both the positions of conformational equilibria (e.g. of OH groups) and the rates of conformational interchange (e.g. nitrogen inversions).

There are many examples in the literature of hydrogen bonding influencing the positions of conformational equilibria in heterocycles. The axial conformations of 5-hydroxy-1,3-dioxan[24] (8) and 5-hydroxy-1,3-dithians (9)[25] are respectively ca 1.0 and 0.5 kcal mol^{-1} more stable than the equatorial. 2,2-Dimethyl-5-hydroxy-1,3-dithian (10) shows a slightly greater axial preference of 0.8 kcal mol^{-1}.[26]

In the piperidine series several clear-cut examples of hydrogen bonding influencing the positions of conformational equilibria have been found. 3-Hydroxypiperidine and its N-methyl derivative show absorption bands in the infrared from both bound and free hydroxyl groups. Both axial and equatorial 3-hydroxy forms (11) are therefore present.[27] The 3-phenyl-3-hydroxypiperidines (12), however, show only bound hydroxyl from the axial conformation. In this case the equatorial preference of the phenyl group is pushing the equilibrium already noted for (11) much more to the side of the axial hydroxyl group, and increasing the free energy difference between the two possible chair conformations.[27]

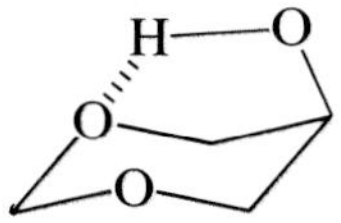

(8) 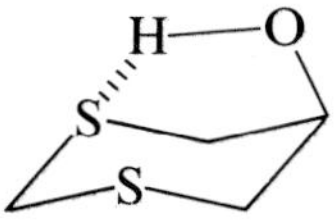   (9) 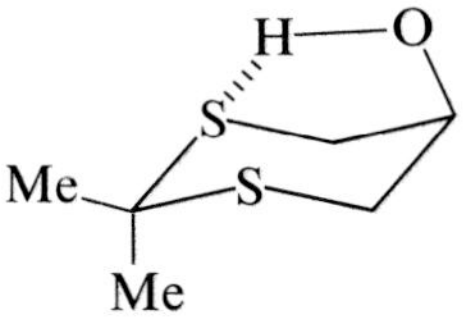   (10)

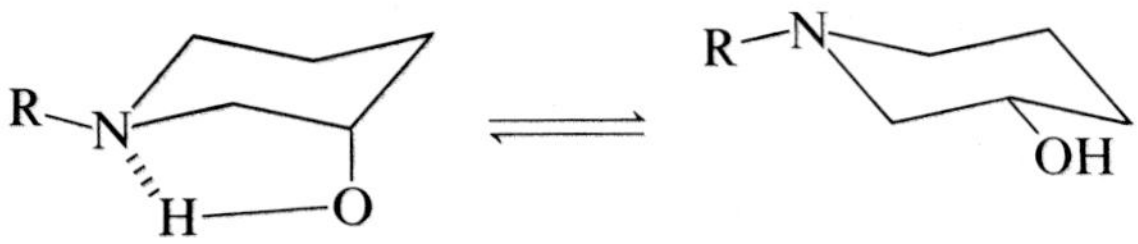

(11a)   (11b) 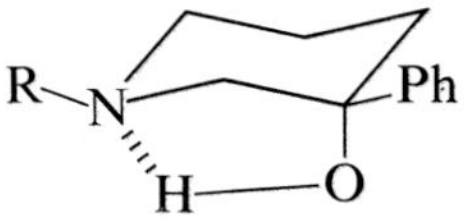   (12)

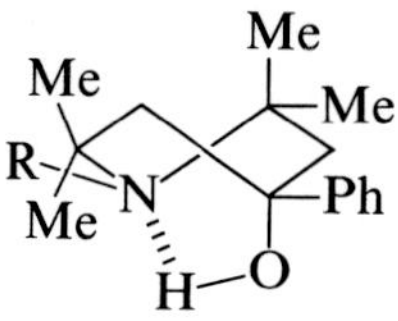

(13) 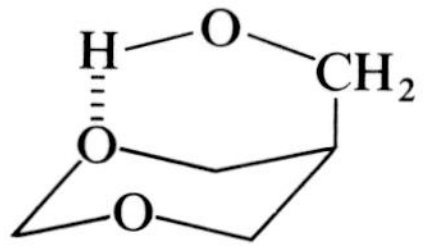   (14)

Although 1-methyl-4-hydroxypiperidine shows no evidence of hydrogen bonding in its infrared spectrum, the 1,2,2,6,6-pentamethyl-4-phenyl-4-hydroxy derivative does, and presumably exists in a twist conformation (13).[28] This was one of the earliest authenticated examples of a non–chair conformation of a six-membered ring being preferred.

It is not always certain that a hydrogen bond will be present when it is expected. For example, the anticipated hydrogen bond between an axial 5-hydroxymethyl group and the ring oxygen atoms in 1,3-dioxan derivatives is not observed (14).[29] Studies of hydrogen bonding in related compounds in the dioxan and tetrahydropyran series[30] suggest that hydrogen bond formation in this type of situation is prevented by a combination of dipolar and bond-eclipsing forces.

Methods of Investigation

The main methods employed for investigating alicyclic conformations can all be applied to the study of heterocyclic systems. However, the presence of heteroatoms in the ring means that many new methods can be employed and that the scope of many other methods is considerably improved. For instance, the wider range of chemical shifts present in heterocyclic compounds dramatically increases the power of NMR spectroscopy.

Equilibration

Chemical equilibration of model compounds has been used extensively in conformational studies and the scope and limitation of the technique are now well understood.[31] Typically in alicyclic systems equilibria are set up by epimerization at a centre next to a carbonyl group. The scope for epimerization improves in saturated heterocyclic compounds. A typical example from the heterocyclic field is the acid-catalysed equilibration of 1,3-dioxans.[32] The 1,3-dioxan ring is an acetal (or ketal), which can be reversibly opened upon treatment with a suitable acid catalyst. As a model

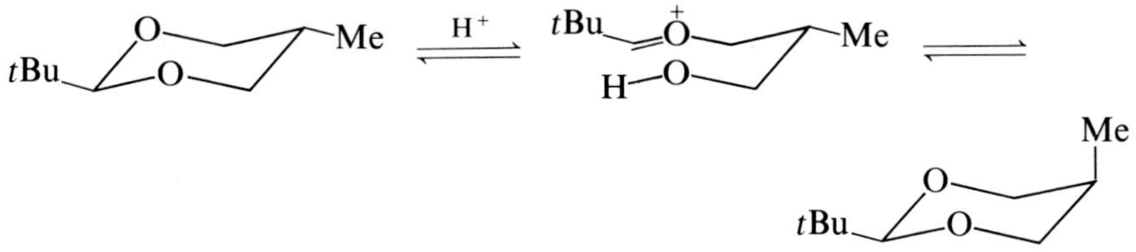

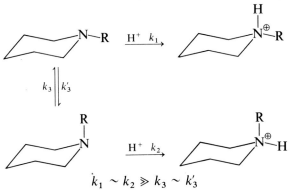

Fig. 2.9   Acid-catalysed chemical equilibrium in a 1,3-dioxan as a model for the conformational equilibrium.

for the conformational equilibrium of a 5-methyl group, 2-*t*-butyl-5-methyl-1,3-dioxan was studied (Fig. 2.9). Making the required assumptions for interpretation of equilibration data, that the bulky *t*-butyl has no effect on the ring, either sterically or electronically, then the chemical equilibrium in Fig. 2.9 is a good model for the corresponding conformational equilibrium. The acetal system in the 1,3-dioxan ring is ideally set up for this acid-catalysed equilibration and similar experiments have been performed on related systems, notably 1,3-oxathians and 1,3-dithians (Chapter 6).

Trapping of conformations

If the rate of a non-reversible chemical reaction or other suitable trapping technique is very much faster than the rate of a conformational interconversion, the rapid reaction may be used to trap and identify the components in the slower conformational equilibrium. This technique is neatly exemplified in the alicyclic field by Anet's technique of rapidly condensing hot cyclohexane (800°) onto the window of an infrared cell cooled to 20 K. The

Fig. 2.10   Trapping of piperidine conformations by rapid protonation (Refs 34 and 35).

spectrum of the condensate allowed estimation of the amount of twist form present in cyclohexane at elevated temperatures.[33]

This method is also demonstrated in the heterocyclic field by Robinson's and McKenna's work on the kinetically controlled protonation of piperidines.[34, 35] A piperidine derivative in which nitrogen inversion is occurring, is diffused into strong acid, either from the vapour phase or from a dilute solution in a solvent such as cyclohexane (Fig. 2.10). By demonstrating that the protonation was non-reversible under the experimental conditions it was possible to conclude that the proportions of salts found in the acid solutions were identical to the proportions of conformations in the other phase. This immediately enables a measurement of the conformational free energy differences in these piperidine derivatives. Other trapping reactions on piperidines using nitrenes give a similar result.[35]

Nuclear magnetic resonance spectroscopy

Nuclear magnetic resonance spectroscopy is undoubtedly the most powerful technique commonly used in heterocyclic conformational analysis. Up to the time of writing, the nuclei most widely used have been ^{1}H and ^{13}C, but in future the ability to use spectrometers capable of giving, for example, ^{15}N, ^{17}O, and ^{33}S spectra in natural abundance, may prove of great value to the subject.

The presence of electronegative heteroatoms gives heterocycles a much wider range of chemical shifts in both ^{1}H and ^{13}C spectra than is possible for alicyclic compounds. This increases the power of NMR methods. First, it gives spectra that are far more readily interpreted, and secondly it minimizes the problems due to second-order splitting patterns in ^{1}H spectra. In many cases the ^{1}H spectra of heterocyclic compounds at, say, 100 MHz are amenable to a first-order treatment of coupling constants.

The Karplus equation relates vicinal ^{1}H coupling constants to dihedral angle in the ethane molecule:

$$J = A \cos 2\theta + B \cos \theta + C \qquad \text{eq. 2.2}$$

and is often approximated as:

$$J = A \cos^2 \theta \qquad \text{eq. 2.3}$$

Although the original derivation of this equation by Karplus was from a valence bond treatment of the ethane molecule it is now widely accepted that it is a good representation of the variation of the coupling constant J with dihedral angle θ in any CH—CH fragment provided that the appropriate constants are employed for the molecule in question.

For the normal *gauche* and *trans* relationships found in molecules it is clear from eq. 2.3 that a *trans* vicinal coupling, $\theta = 180°$, is greater than a *gauche* vicinal coupling, $\theta = 60°$. Thus in a six-membered ring axial–axial vicinal couplings will be larger (8–15 Hz) than equatorial–equatorial or equatorial–axial couplings (0–6 Hz) enabling stereochemical relationships to be readily defined.

The principal difficulty in applying the Karplus equation quantitatively in heterocyclic systems is that the value of the constants in eqs 2.2 and 2.3 are often unknown. The constant A in eq. 2.3 varies with the electronegativity of the substituent on the C–C bond. Lambert solved this problem by taking ratios of coupling constants (R value) for the couplings in the same vicinal fragment, thus eliminating the constant A.[36]

$$R = \frac{J_{trans}}{J_{cis}} \qquad\qquad \text{eq. 2.4}$$

For rapidly inverting six-membered rings R values near 2.0 are indicative of an almost perfect chair conformation, whilst lower R values nearer to 1.0 suggest the presence of flattened chairs or even non-chair conformations. Highly puckered chairs have $R > 2$. The R value method, however, has severe limitations when applied to rings other than six-membered.[37]

Buys has shown that for many six-membered rings the internal ring dihedral angle (ω) can be related to R by the eq. 2.5.[38]

$$R = \frac{J_{trans}}{J_{cis}} = \frac{(3 - 2\cos^2\omega)}{4\cos^2\omega} \qquad\qquad \text{eq. 2.5}$$

This appears to be a good way for determination of dihedral angles in saturated heterocyclic rings. Values of dihedral angles obtained by this method generally agree to within 2° of those found by other methods.[39]

Analysis of NMR lineshapes for systems undergoing chemical exchange, a technique sometimes referred to as Dynamic Nuclear Magnetic Resonance, yields rate constants for the exchange process. In practice the rate constant for the exchange process must be round about the value of the chemical shift difference measured in Hz between the exchanging sites. Generally, to bring the rate of exchange into the observable range, the temperature of the sample is varied. Accessible temperatures range from ca 100 K to ca 500 K, giving a variation in free energies of activation of from ca 5 kcal mol^{-1} to ca 25 kcal mol^{-1}. This technique has proved invaluable in the measurement of rates of conformational interchange and in determining the proportions of various conformations present at the slower exchange rates (slow exchange limit) when the spectrometer "sees" all of the slowly interconverting conformations.

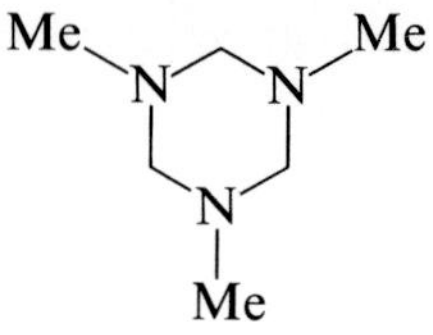

(15)

The method is exemplified by studies of 1,3,5-trimethylhexahydro-1,3,5-triazine (15).[40,41] At room temperature the ^{1}H spectrum consists of two singlets in the ratio 2:3 arising from the ring methylene and N-methyl hydrogens. This spectrum is consistent with rapid inversion of the ring and nitrogen atoms, giving rise to a time-averaged spectrum. As the temperature is lowered the ring methylene resonance broadens, and by $-50°$ has split into an AB quartet. The coalescence temperature (T_c) is at $-5°$. This observation is consistent with slowing down of ring inversion whilst nitrogen inversion remains rapid and gives a value for the free energy of activation $\Delta G^{\ddagger}$ for ring inversion of ca 13 kcal mol^{-1} (Fig. 2.11).[40]

If the sample is cooled further, more changes take place. Below $-100°$ the resonances broaden and then sharpen with T_c at $-123°$ and $\Delta G^{\ddagger}$ ca 7.2 kcal mol^{-1}. At $-144°$ the spectrum shows two N-methyl peaks in the ratio 1:2 and two quartets for the ring methylenes also in the ratio 1:2. These low temperature changes arise from the slowing down of nitrogen inversion leaving the predominant conformation (16) visible in the slow exchange limit spectrum. This conformation has one axial and two equatorial methyl groups.[41a] Analogous changes are observed in the ^{13}C spectra.[41b] Whilst most studies of rates of conformational interchange and barriers to confor-

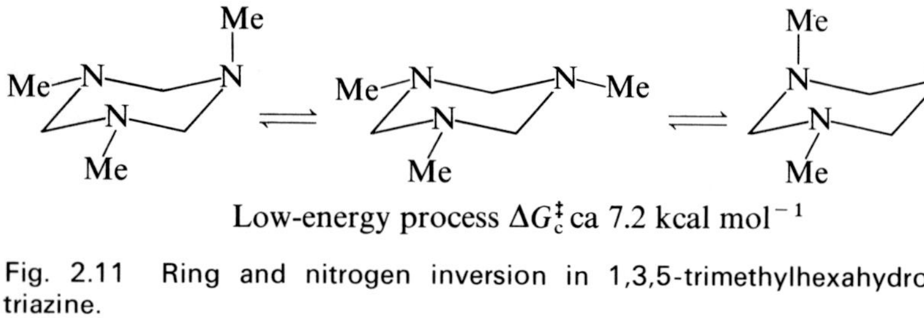

High-energy process $\Delta G_c^{\ddagger}$ ca 13 kcal mol^{-1}

Low-energy process $\Delta G_c^{\ddagger}$ ca 7.2 kcal mol^{-1}

Fig. 2.11   Ring and nitrogen inversion in 1,3,5-trimethylhexahydro-1,3,5-triazine.

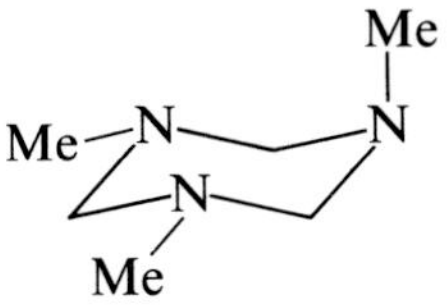

(16)

mational processes have been of systems with equally populated sites, it is also possible to study those phenomena with very unequal site populations. Equations developed by Anet et al.[42] give

$$v_{(\frac{1}{2}max)} = P \times \Delta v \qquad \text{eq. 2.6*}$$

$$k = 2\pi \, \Delta v \qquad \text{eq. 2.7*}$$

Equation 2.6 gives the population of a minor form from the maximum line broadening provided that the chemical shift difference is known. Equation 2.7 gives the rate constant for exchange at maximum broadening and hence the free energy of activation.

These equations were first used in the study of 1,2,2,6-tetramethyl-piperidine Fig. 2.12.[42] It was possible to deduce that $\Delta G^{\ddagger}$ for nitrogen inversion was $11.0 \pm 0.3 \, \text{kcal mol}^{-1}$ (eq $\rightarrow$ ax) and that the equatorial N-methyl group is $1.9 \pm 0.2 \, \text{kcal mol}^{-1}$ more stable than axial at 213 K. The latter result was confirmed by conformational trapping using kinetic protonation (see p. 33).

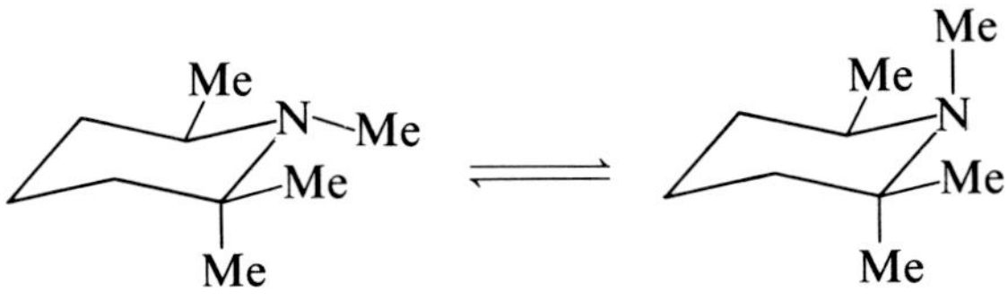

Fig. 2.12   Nitrogen inversion in 1,2,2,6-tetramethylpiperidine.

These equations have subsequently been used by other workers and have proved extremely valuable especially for ^{13}C studies of low barriers not accessible to ^1H NMR spectrometers.

The widespread use of shift reagents to simplify NMR spectra merits a note of caution when applied to conformational studies. It has been shown in several cases that complexation of the shift reagent with the substrate

---

* $v_{(\frac{1}{2}max)}$ is the maximum broadening at half height of the signal; P is the population of the minor form; Δv is the chemical shift difference in Hz, and k is the rate constant in s^{-1} for the temperature of maximum broadening in the direction minor to major. These equations are applicable for conformer ratios greater than 10:1 and can be obtained mathematically from the Gutowsky–Holm lineshape equation for two sites.

can perturb conformational equilibria. In particular, piperidines show a differential complexation of equatorial and axial lone pairs with cobalt acetyl acetonate,[43] changes in coupling constants are observed when some 2-alkyl-2-oxo-1,3,2-dioxaphosphorinans are treated with Eu(fod)$_3$,[44] and conformational changes place when shift reagents interact with cyclohexyl carbonitrile.[45]

Carbon-13 NMR spectra are particularly useful in the study of conformational problems. The major advantages of ^{13}C spectra over ^{1}H spectra are the tremendous spread of chemical shifts available (up to 600 ppm for ^{13}C vs ca 10 for ^{1}H) and the absence of coupling constants in noise-decoupled spectra. These generally render the signal of each carbon atom in the compound visible. The major disadvantage, the low natural abundance of ^{13}C, is nowadays largely overcome by the use of Fourier Transform instrumentation.

One major feature of ^{13}C NMR spectra that is particularly useful in conformational studies is the almost linear additivity of substituent effects upon ^{13}C chemical shifts. This makes it possible to construct schemes for predicting the chemical shifts of well-defined conformations.

The chemical shifts obey the equation:

$$\Delta_{xn} = \Delta_{xp} + \sum_1^y E_x^y + k \qquad \text{eq. 2.8}$$

where Δ_{xn} is the chemical shift of the xth ring carbon atom in the nth compound of a series, Δ_{xp} is the shift of the same carbon atom in the parent unsubstituted compound, and the terms E_x^y are the y effects of substituents on the chemical shift of the xth carbon atom. The constant k is usually arbitrarily set to 0. The usual technique is to measure the shifts for ring carbon atoms in a wide variety of derivatives of a given ring in order to obtain an over-determined set of linear equations based on eq. 2.8. The statistical technique of multiple linear regression analysis is then used to determine the best fit values of the parameters E_x^y.

This technique has been used *inter alia* in an examination of the 1,3-dioxan,[46] 1,3-dithian,[47] and dihydro-1,2-oxazine[48] rings. It has proved particularly useful in demonstrating conformational abnormalities. For example, it has shown the presence of twist conformations in certain derivatives of 1,3-dioxan that would contain exceptional strain in chair conformations,[46] and has shown the presence of low-energy twist conformations in the 1,3-dithian series.[47] Its use is not restricted to saturated systems and it has been employed in investigations of the cyclohexane analogue, dihydro-1,2-oxazine.[48] The further potential of this technique awaits development.

The advent of high field superconducting magnets in Fourier Transform instruments will make studies of nuclei such as ^{15}N and ^{17}O much easier.

There is obviously great potential in applying some of the above techniques to the study of these nuclei.

Microwave spectroscopy

For simple molecules microwave spectroscopy can give highly detailed conformational information.[49] Saturated heterocyclic molecules might seem ideal candidates for study by this technique since they generally possess permanent dipole moments, a basic requirement for microwave spectra. However, most six-membered (or larger) rings give microwave spectra that are too complex for the extensive line assignment and computation that is required. Consequently, the number of six-membered rings that have been successfully studied is very small. For smaller rings the technique has been successfully applied.

In four-membered rings microwave methods have demonstrated the essentially planar nature of trimethylene oxide, where the barrier to ring inversion is below the ground vibrational level, and have permitted determination of the ring inversion barriers in several other systems (Chapter 3).[49] Several five-membered rings have also been investigated.[49]

Piperidine succumbed to a prolonged, time-consuming attack by microwave methods,[50] and Kewley et al. have reported results on some other common six-membered rings.[51]

Electron diffraction

Although in certain circumstances electron diffraction is an extremely powerful method for conformational studies,[52] it is less useful in the case of heterocyclic compounds. This is because the type of heterocyclic compound with which we are concerned in this book contains many similar types of interatomic distances which prove difficult to disentangle in the radial distribution curve. Nonetheless, electron diffraction studies on five-membered rings have shown, for example, that pseudorotation in tetrahydrofuran is essentially free,[53] but that there is an appreciable pseudorotation barrier in the analogous compounds tetrahydrothiophene and tetrahydroselenophene, and that the preferred conformations of these rings have C_2 symmetry.[54,55]

Ultrasonic relaxation

In recent years there has been a reawakening of interest in ultrasonic relaxation techniques.[56] Although there is no special advantage in this technique for heterocycles as compared to other systems a few words of explanation and caution are in order here.

Results from ultrasonic relaxation must be treated with caution for two main reasons. First there is no means of exactly ascertaining the origin of the relaxation observed. It is always conceivable that the results may be due to some minor impurity present in the sample rather than the process being sought. A noted example of this is the report of relaxation in cyclohexene said to arise from interconversion between chair and boat forms.[57] This result could not be reproduced by other workers and has since been attributed to ca 20% cyclohexane in the original sample.[58, 59] Secondly, although activation energies for conformational processes may now be measured with some confidence by this technique, the magnitudes of conformational energy differences are measured with a much greater uncertainty.[60] These barriers are derived for the process in the direction least stable conformation to most stable conformation.[61] This point has not been widely appreciated in the past leading to some ill-founded criticism of the technique.[62] Amongst the major workers in this field has been Wyn-Jones who has studied 1,3-dioxans,[63] cyclic sulphites,[64] and piperidines.[65]

Dipole moments

The presence of polar heteroatoms in heterocyclic molecules renders measurements of molecular dipole moments a feasible technique for conformational analysis. Whether or not a molecule will possess an overall dipole moment will depend upon the symmetry of its conformations. Conformations that lack any symmetry, or fall into point groups C_n C_{nv} or C_s, must possess permanent dipole moments. Conformations in other point groups can have no dipole moment. The time average of the conformations thus determines the observed dipole moment.

Conformational results derived from dipole moment studies have been used fairly extensively in the past but must be treated with some caution. Too often the dipole moment technique has given results that have later been shown to be quantitatively or even qualitatively incorrect.

The technique and its inherent dangers are best illustrated in the context of an actual example.[66] To determine the position of the conformational equilibrium and hence the free energy difference at the nitrogen atom in N-alkylpiperidines the dipole moments of 4-p-nitro and 4-p-chlorophenyl derivatives (17) were measured. The following assumptions, each of which is subject to some doubt, were then explicitly made: (a) a specific piperidine ring geometry was assumed; (b) the direction of the component dipole moments in each conformation was assumed; (c) the non-interference of the 4-aryl group with the conformational equilibrium was assumed; (d) the

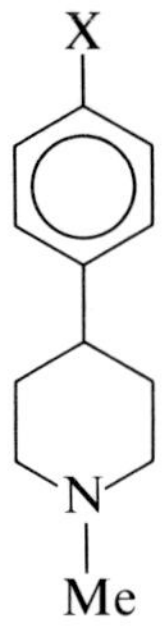

X = NO_2 or Cl

(17)

aryl groups were assumed to be exclusively equatorial; (e) the component dipole moments were then assumed to be those in 1-methylpiperidine and p-nitro or p-chlorophenylcyclohexane. Vector addition of the dipole moments under these sets of assumptions gave the two dipole moments for equatorial and axial N-alkyl groups. The mole fractions of each conformation were then calculated from the observed value using eq. 2.9

$$\mu_{obs}^2 = n_e \mu_e^2 + (1 - n_e)\mu_a^2 \qquad \text{eq. 2.9}$$

Since it is the polarizability that is the linearly additive molar quantity, and since the polarizability is proportional to μ^2, eq. 2.6 contains μ^2 terms and not μ.

Qualitatively, the answer obtained from this experiment is correct. It is found that the equatorial form is more stable by ca 0.65 kcal mol^{-1} for N-methyl. Quantitatively, however, the result is at serious variance with the now generally accepted value of 2.7 kcal mol^{-1}.[34,35,42]

It is clear that as the proportion of the minor conformation decreases so does its influence in eq. 2.9. Dipole moment methods are therefore of little use for strongly biassed equilibria (say, $K > 4$, i.e. $< 20\%$ minor component).

Vibrational spectroscopy

The two complementary vibrational spectroscopic methods—infrared and Raman spectroscopy—have several definite roles in heterocyclic conformational analysis. Infrared spectra arise when a molecular vibration leads to a change in dipole moment. Raman spectra arise when the vibration leads to a change in molecular polarizability.

Infrared spectra are of use in several different ways. First, as we have discussed above they are used to record the presence or absence of hydrogen bonds. Secondly, they can be used in the qualitative detection of certain conformations e.g. the so-called Bohlmann bands (see below). Thirdly, they may be used in the quantitative evaluation of the relative amounts of conformations which give separate and identifiable bands. Finally, far infrared spectra can give estimates of barrier heights for certain conformational processes.

The Bohlmann bands which occur at 2700–2800 cm^{-1} are associated with at least two axial hydrogens antiperiplanar to a pair of electrons on nitrogen.[67,68] Their presence in a spectrum therefore implies that the N-substituent in a six-membered ring is equatorial (18).

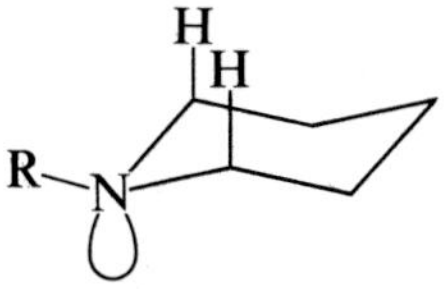

(18)

Since in infrared spectroscopy the transitions are virtually instantaneous in comparison with the lifetime of a given conformation,* separate bands can be seen and sometimes identified for particular conformations. The most elegant demonstration of this technique is to be found in the variable-temperature infrared study of piperidine by Baldock and Katritzky.[69] The gas phase spectrum shows two resolved bands in the N—H first overtone region. These bands were assigned to the N—H axial and equatorial conformations by comparison of their contours with those predicted by theory. Temperature variation of the intensity ratios then allowed an estimate of the rate at which the equilibrium constant varied with temperature leading to a value of ΔH. The equatorial conformation is the more stable by 0.5 ± 0.1 kcal mol^{-1}. This is a rare example of a case where an enthalpy difference is known more accurately than a free energy difference. The extinction coefficients of the two bands are not known and cannot be assumed to be equal. Thus the equilibrium constants K can never be measured by this technique, although the rate of variation of K with temperature is readily determined.

The use of far infrared spectra to determine conformational parameters is illustrated by the measurement of the ring inversion barrier in dioxene (19).

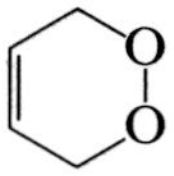

(19)

* The opposite holds for NMR spectra which in most instances are time averaged over all conformations.

Both NMR and IR methods agree on a barrier of 7.62 ± 0.15 kcal mol^{-1} for this process.[70]

The combined use of Raman and infrared spectral data can provide valuable symmetry information about conformations. Since the selection rules differ for the two forms of spectroscopy, some bands are active in both and some in just one of these spectroscopies. Combining this information gives the molecular symmetry. This method has been used by Lambert as a means of differentiating between chair and twist conformations of six-membered rings,[71a] by Katritzky *et al.* for hexahydro-1,2,4,5-tetrazines,[71b] and clearly is of potential value for heterocyclic compounds.

When a complete assignment of the vibrational frequencies has been made it is frequently possible to use this data to construct a vibrational force field for the whole molecule. This has possibilities for use in conformational analysis.[72]

Photoelectron spectroscopy

In photoelectron spectroscopy the molecules of the compound under investigation are bombarded by a beam of photons. In practice this is typically the helium line at 58 400 pm $\equiv$ 21.21 eV. This causes the molecule to eject electrons with characteristic kinetic energies. The kinetic energy of an ejected electron corresponds to the difference between the relevant ionization potential and the photon energy. The overall spectrum thus corresponds to a series of ionization potentials for the various electrons in the molecule. As with the optical spectroscopies these electronic transitions are very much more rapid than conformational interchanges and thus one obtains the spectra of all the conformations present superimposed on each other.

In heterocyclic conformational analysis photoelectron spectra have proved particularly useful in the study of hydrazine derivatives. It turns out that the lone pair orbitals on nitrogen in hydrazine derivatives interact to form a symmetric and an antisymmetric combination.[73] The relative energies of these orbitals vary with the dihedral angle between the lone pairs, therefore the splitting between the two ionization potentials will be conformation dependent. This splitting has been used to good effect by Nelsen,[12,74] and also by Rademacher[75] in investigations of cyclic hydrazines.

References

1. L. Pauling. (1960). "The Nature of the Chemical Bond", Cornell University Press, Ithaca.
2. J. B. Lambert and S. I. Featherman. (1975). *Chem. Rev.* **75**, 611.

3. G. M. Kellie and F. G. Riddell. (1974). *In* "Topics in Stereochemistry", Vol. 8, (Eds Eliel and Allinger), 225. Wiley Interscience, London, New York.
4. R. H. Hunt, R. A. Leacock, C. W. Peters and K. T. Hecht. (1965). *J. Chem. Phys.* **42**, 1931.
5. W. H. Fink and L. C. Allen. (1967). *J. Chem. Phys.* **46**, 2261, 2276; L. Pedersen and K. Morokuma. (1967). *J. Chem. Phys.* **46**, 3941.
6. F. G. Riddell, E. S. Turner, D. W. H. Rankin and M. R. Todd. (1979). *J.C.S. Chem. Comm.* 72.
7. R. P. Lattimer and M. D. Harmony. (1970). *J. Chem. Phys.* **53**, 4575.
8. K. Wulz, H. A. Brune and W. Hetz. (1970). *Tetrahedron.* **26**, 3; (1971). **27**, 3629.
9. G. Wood and R. M. Srivastava. (1971). *Tetrahedron Lett.* 2937.
10. F. G. Riddell and D. R. Slim. Unpublished results calculated from data by P. R. Groth. (1967). *Acta Chem. Scand.* **21**, 2608, 2631, 2695.
11. F. G. Riddell, P. Murray-Rust and J. Murray-Rust. (1974). *Tetrahedron* **30**, 1087.
12. S. F. Nelson. (1978). *Acc. Chem. Res.* **11**, 14.
13. R. M. Moriarty. (1974). *In* "Topics in Stereochemistry", Vol. 8, (Eds Eliel and Allinger), 271. Wiley Interscience, New York, London.
14. S. Wolfe. (1972). *Acc. Chem. Res.* **5**, 102.
15. R. J. Abraham and P. Loftus. (1974). *J.C.S. Chem. Comm.* 180.
16. N. S. Zefirov, S. V. Rogozina, E. H. Kurkutova, A. V. Goncharev and N. V. Belov. *J.C.S. Chem. Comm.* (1974). 260; N. S. Zefirov and S. V. Rogozina, (1974). *Tetrahedron* **30**, 2345.
17. N. S. Zefirov, L. G. Gurvich, A. S. Shashkov, M. Z. Krimer and E. A. Vorob'eva. (1976). *Tetrahedron* **32**, 1211.
18. E. L. Eliel and E. Juaristi. (1978). *J. Am. Chem. Soc.* **100**, 6114.
19. M. L. Stein, G. Chiurdoglu, R. Ottinger, J. Reisse and H. Christol. (1971). *Tetrahedron* **27**, 411; R. D. Stolow, D. I. Lewis and P. A. D'Angelo. (1970). *Tetrahedron* **26**, 5831.
20. C. Romers, C. Altona, H. R. Buys and E. Havinga. (1969). *In* "Topics in Stereochemistry", Vol. 4, (Eds Eliel and Allinger), 39. Wiley Interscience, New York, London.
21. S. David, O. Eisenstein, W. J. Hehre, L. Salem and R. Hoffmann. (1973). *J. Am. Chem. Soc.* **95**, 3806.
22. R. U. Lemieux. (1971). *Pure Appl. Chem.* **25**, 527.
23a. W. F. Bailey and E. L. Eliel. (1974). *J. Am. Chem. Soc.* **96**, 1798.
23b. R. U. Lemieux. (1971). *Pure Appl. Chem.* **25**, 527.
24. N. Bagget, M. A. Bukhari, A. B. Foster, J. Lehmann and J. M. Weber. (1963). *J. Chem. Soc.* 4157.
25. A. Lüttringhaus, S. Kabuss, H. Prinzbach and F. Langbircher. (1962). *Annalen* **653**, 195.
26. R. J. Abraham and W. A. Thomas. (1965). *J. Chem. Soc.* 335.
27. G. Hite, E. E. Smissman and R. West. (1960). *J. Am. Chem. Soc.* **82**, 1207.
28. R. E. Lyle. (1957). *J. Org. Chem.* **22**, 1280.
29. E. L. Eliel and M. K. Kaloustian. *Chem. Comm.* (1970). 290; R. Dratler and P. Laszlo. (1970). *Tetrahedron Lett.* 2607.
30. E. L. Eliel and H. D. Banks, *J. Am. Chem. Soc.* (1970), **92**, 4730; (1972). **94**, 171.
31. F. G. Riddell. (1974). *In* "Internal Rotation in Molecules", (Ed. W. J. Orville Thomas), 19. Wiley, London.
32. F. G. Riddell and M. J. T. Robinson. (1967). *Tetrahedron* **23**, 3417; E. L. Eliel and M. C. Knoeber. (1968). *J. Am. Chem. Soc.* **90**, 3444.

33. M. Squillacote, R. S. Sheridan, O. L. Chapman and F. A. L. Anet. (1975). *J. Am. Chem. Soc.* **97**, 3244.
34. P. J. Crowley, M. J. T. Robinson and M. G. Ward, *J.C.S. Chem. Comm.* (1974). 825; K. W. Baldry and M. J. T. Robinson. (1975). *Tetrahedron* **31**, 2621; P. J. Crowley, M. J. T. Robinson and M. G. Ward. (1977). *Tetrahedron* **33**, 915.
35. D. C. Appleton, J. McKenna, J. M. McKenna, L. B. Simms and A. R. Walley. (1976). *J. Am. Chem. Soc.* **98**, 292.
36. J. B. Lambert. (1967). *J. Am. Chem. Soc.* **89**, 1836; (1971). *Acc. Chem. Res.* **4**, 87.
37. J. B. Lambert, J. J. Papay, E. S. Magyar and M. K. Neuberg. (1973). *J. Am. Chem. Soc.* **95**, 4459; J. B. Lambert, J. J. Papay, S. A. Khan, K. A. Kappauf and E. S. Magyar. (1974). *J. Am. Chem. Soc.* **96**, 6112.
38. H. R. Buys. (1970). *Rec. Trav. Chim.* **89**, 1244.
39. H. R. Buys. (1970). *Rec. Trav. Chim.* **89**, 1253.
40. J. M. Lehn, F. G. Riddell, B. J. Price and I. O. Sutherland. (1967). *J. Chem. Soc. (B)* 387.
41a. C. H. Bushweller, M. Z. Lourandos and J. A. Brunelle. (1974). *J. Am. Chem. Soc.* **96**, 1591.
41b. V. J. Baker, I. J. Ferguson, A. R. Katritzky, R. C. Patel and S. Rahimi-Rastgoo. (1978). *J.C.S. Perkin II* 377.
42. F. A. L. Anet, I. Yavari, I. J. Ferguson, A. R. Katritzky, M. Moreno-Mañas and M. J. T. Robinson. (1976). *Chem. Comm.* 399.
43. I. D. Blackburne, A. R. Katritzky and Y. Takeuchi. (1974). *J. Am. Chem. Soc.* **96**, 682.
44. W. G. Bentrude, H.-W. Tan and K. C. Yee. (1972). *J. Am. Chem. Soc.* **94**, 3264; and A. J. Dale. (1972). *Acta Chem. Scand.* **26**, 2985.
45. D. J. Raber, M. D. Johnston and M. A. Schwalke. (1977). *J. Am. Chem. Soc.* **99**, 7671.
46. G. M. Kellie and F. G. Riddell. (1971). *J. Chem. Soc. (B)* 1030.
47. E. L. Eliel, V. S. Rao and F. G. Riddell. (1976). *J. Am. Chem. Soc.* **98**, 3583.
48. H. Labaziewicz, F. G. Riddell and B. G. Sayer. (1977). *J.C.S. Perkin II* 619.
49. N. L. Owen. (1972). *In* "Internal Rotation in Molecules", (Ed. W. J. Orville Thomas), 157. Wiley, London.
50. P. J. Buckley, C. C. Costain and J. E. Parkin. (1968). *Chem. Comm.* 668.
51. See for example S. C. Dass and R. Kewley. (1974). *Canad. J. Chem.* **52**, 434.
52. A. H. Clark. (1972). *In* "Internal Rotation in Molecules", (Ed. W. J. Orville Thomas), 325. Wiley, London.
53. A. Almenningen, H. M. Seip and T. Willadsen. (1969). *Acta Chem. Scand.* **23**, 2748.
54. Z. Nahlovska, B. Nahlovsky and H. M. Seip. (1969). *Acta Chem. Scand.* **23**, 3534.
55. Z. Nahlovska, B. Nahlovsky and H. M. Seip. (1970). *Acta Chem. Scand.* **24**, 1903.
56a. S. M. Walker. (1972). *In* "Internal Rotation in Molecules", (Ed. W. J. Orville Thomas), 286. Wiley, London.
56b. E. Wyn-Jones and R. A. Pethrick. (1970). *In* "Topics in Stereochemistry," Vol. 5, (Eds Eliel and Allinger), 205. Wiley Interscience, New York.
57. J. Karpovitch. (1954). *J. Chem. Phys.* **22**, 1767.
58. K. R. Crook and E. Wyn-Jones. (1971). *Trans. Faraday Soc.* **67**, 660.
59. S. V. Subrahmanyam and J. E. Piercy. (1965). *J. Chem. Phys.* **42**, 1845.
60. P. J. Heywood, J. E. Rossing and E. Wyn-Jones. (1975). *Adv. Molec. Relax. Proc.* **6**, 255.
61. E. Wyn-Jones. (1971). *Tetrahedron Lett.* 907.

62.   G. Wood, J. M. McIntosh and M. H. Miskow. (1970). *Tetrahedron Lett.* 4895.
63.   G. Eccleston, E. Wyn-Jones and W. J. Orville Thomas. (1971). *J. Chem. Soc.* (*B*) 1551, 2469.
64.   G. Eccleston, E. Wyn-Jones, P. C. Hamblin and R. F. M. White. (1970). *Trans. Faraday Soc.* **66**, 310.
65.   V. M. Gittins, P. J. Heywood and E. Wyn-Jones. (1975). *J.C.S. Perkin II* 1642.
66.   R. A. Y. Jones, A. R. Katritzky, A. C. Richards and R. J. Wyatt. (1970). *J. Chem. Soc.* (*B*) 122.
67.   F. Bohlmann. (1958). *Chem. Ber.* **91**, 2157.
68.   G. W. Gribble and R. B. Nelson. (1973). *J. Org. Chem.* **38**, 2831.
69.   R. W. Baldock and A. R. Katritzky. *Tetrahedron Lett.* (1968). 1159; *J. Chem. Soc.* (*B*) (1968). 1470.
70.   R. H. Larkin and R. C. Lord. (1973). *J. Am. Chem. Soc.* **95**, 5129.
71a. D. S. Bailey and J. B. Lambert. (1973). *J. Org. Chem.* **38**, 134.
71b. R. A. Y. Jones, A. R. Katritzky, A. R. Martin, D. L. Ostercamp, A. C. Richards and J. M. Sullivan. (1974). *J.C.S. Perkin II* 948.
72.   See for example: O. H. Ellestad, P. Klaboe and G. Hagen. *Spectrochim. Acta* (1971). **27A**, 1025; (1972). **28A**, 137; O. H. Ellestad, P. Klaboe, G. Hagen and T. Stroyer-Hansen. *Spectrochim. Acta* (1972). **28A**, 149.
73.   R. Hoffmann. (1971). *Acc. Chem. Res.* **4**, 1.
74.   S. F. Nelsen and J. M. Buschek. *J. Am. Chem. Soc.* (1974). **96**, 6982, 6987, 7930; (1973). **95**, 2011, 2013.
75.   P. Rademacher. (1975). *Chem. Ber.* **108**, 1548; P. Rademacher and H. P. Koopmann. (1975). *Chem. Ber.* **108**, 1557.

3
Four- and Five-Membered Rings

Four-Membered Rings

Introduction

The conformational analysis of four-membered rings, has been the Cinderella of the subject for a long time. Conceptually it is very simple, but on the experimental front it has attracted few organic chemists. As might perhaps have been expected, the bulk of the experimental conformational information on heterocyclic four-membered rings comes from microwave spectra. Microwave spectroscopy is a particularly appropriate technique for these compounds since they have low molecular weight and possess the essential electrical dipole moment. Thus, very accurate structural and energetic results are available for most of the parent heterocyclic four-membered rings. Infrared and Raman spectroscopy have also provided much valuable information.

Cyclobutane

Cyclobutane itself cannot be studied by microwave spectroscopy as it has no permanent dipole moment, moreover the ring puckering (or out-of-plane) vibration is infrared inactive as it entails no change in dipole moment.*

* See, however, J. M. Stone and I. M. Mills. (1970). *Molec. Phys.* **18**, 631.

Fortunately, however, this vibration is Raman active as it involves a change in molecular polarizability, and thus cyclobutane is an excellent example of how Raman spectroscopy has allowed a study of an otherwise experimentally intractable problem.

Cyclobutane was thought to be a planar ring by von Baeyer in 1885[2] and formed one of the bases of his "strain theory". This view prevailed for a long time despite the fact that Sachse pointed out, as early as 1890, the possibility of a non-planar cyclobutane ring at the same time as he put forward his ideas on the chair and boat forms of cyclohexane.[3] The earliest infrared and Raman work also tended to support the idea of a planar ring.[4]

In 1945 R. P. Bell produced some theoretical quantum mechanical calculations on energy wells involving a fourth-power energy law.[5] In searching for an example to which he might apply his theory he pointed out that the out-of-plane vibrations in four-membered rings should obey a fourth-power law arising from ring valence angle deformation[6] (eq. 3.1).

$$V(\chi) = \alpha\chi^4 \qquad \text{eq. 3.1}$$

In eq. 3.1, χ is a small out-of-plane displacement of one of the ring atoms and α is the force constant. When the energy levels were calculated according to Bell's theory for fourth-power energy wells reasonable agreement was found with the limited data then available.

Equation 3.1 only allows for angle deformation. To allow for torsional changes around the bonds in the ring it is now usual to add a quadratic term, and the expression now becomes:

$$V(\chi) = \alpha\chi^4 - \beta\chi^2 \qquad \text{eq. 3.2}$$

Equation 3.2 is that of a symmetrical double minimum potential energy well (Fig. 3.1). The planar form, advocated by von Baeyer, now becomes an

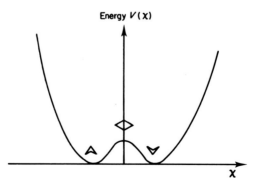

Fig. 3.1   Energy changes during ring inversion in cyclobutane.

energy maximum between two non-planar energy minima in which torsional strain is relieved at the expense of a small amount of angle deformation.

Definitive proof of the non-planar nature of cyclobutane and its derivatives came in 1950, when a Raman study under the improved conditions of resolution then available, showed almost twice as many fundamental vibrations as would be expected to be Raman active in a planar ring.[7]

The ring-puckering vibration leading to non-planarity of cyclobutane may be described by a single coordinate. This may be either, as in eq. 3.2 and Fig. 3.1, the displacement of one ring atom from the plane of the other three or, as is now more generally taken, the dihedral angle between the planes C_1—C_2—C_3 and C_3—C_4—C_1 in Fig. 3.2. In cyclobutane this angle

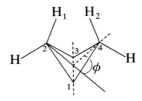

Fig. 3.2   Geometry of cyclobutane.

is found experimentally to be between $23°$ and $35°$.[8,9] It is noteworthy that in the ground state conformation the C—C—C planes do not bisect the H—C—H angles. Atoms H_1 and H_2 in Fig. 3.2 tilt further towards each other with an angle of $4°$. The effect of this is presumably to minimize further the torsional interactions about the C—C bonds at a minimal extra expense in deforming C—C—H bond angles. The barrier to the ring puckering in cyclobutane is found from the Raman measurements to be 1.481 ± 0.014 kcal mol^{-1}.[8]

Because the cyclobutane ring is puckered, a substituent can occupy either of two positions termed axial and equatorial by analogy with cyclohexane (Fig. 3.3). Experimental results show that the equatorial form is generally more stable. Substituted cyclobutanes, therefore, inhabit an

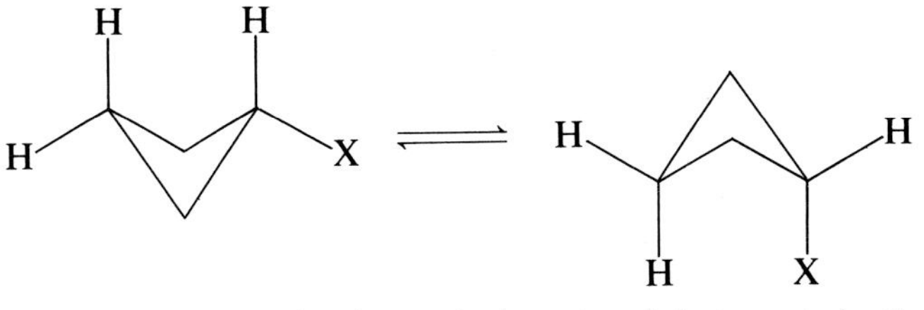

Fig. 3.3   Ring inversion in a substituted cyclobutane derivative.

Energy $V(\chi)$

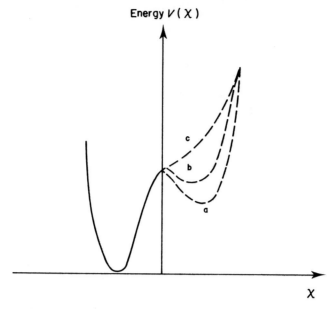

Fig. 3.4   Different forms of asymmetric energy well for a substituted four-membered ring.

unsymmetrical double minimum potential energy well (Fig. 3.4). To allow for this eq. 3.2 is modified by the inclusion of a cubic term:

$$V(\chi) = \alpha\chi^4 - \beta\chi^4 + \gamma\chi^3 \qquad \text{eq. 3.3}$$

The axial form has a finite existence in its own right only if there is one or more allowed vibrational energy levels in the upper potential energy minimum. For the function a in Fig. 3.4, which has a substantial well for the upper minimum, one or more vibrational levels for the axial form may be permitted. For the curves b or c which have either a shallow upper minimum, or merely a point of inflexion, no levels corresponding to a separate axial form may exist. In these latter cases the lower vibrational energy levels correspond to a non-planar ring, whereas the upper levels correspond to an essentially planar molecule.

Oxetane

The microwave spectrum of oxetane (1) shows a very interesting situation that is depicted schematically in Fig. 3.5.[10] The molecule inhabits a double minimum potential energy well with a barrier height of 0.100 kcal mol^{-1}. However, the ground vibrational level is 0.023 kcal mol^{-1} above this

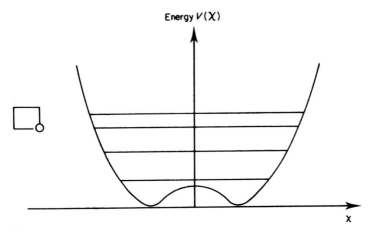

Fig. 3.5 Schematic diagram of potential well and allowed energy levels in oxetane.

barrier. This means that the vibrations of the ring are unhindered by the barrier and are therefore essentially those of a planar molecule.

The substitution of two fluorine atoms for hydrogens in oxetane to give 3,3-difluorooxetane (2) completely removes the barrier,[11] giving a single minimum-energy point in the potential energy well. Two relevant points should be noted. First the barrier in cyclobutane is lowered from 1.48 kcal mol⁻ to 0.69 kcal mol⁻ when the same substitution is performed.[12] Secondly dipolar interactions between the dipole vector sum of the *gem*-difluoro group and the ring oxygen atom will be a minimum in the planar conformation further depressing the barrier.

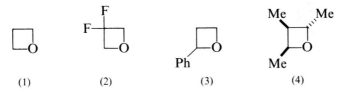

(1)	(2)	(3)	(4)

The ¹H NMR spectra of 2-phenyl,[13] and *r*-2, *cis*-3, *trans*-4-trimethyl oxetane[14] (3) and (4) have been reported and analysed in terms of their ring conformations.

Thietane

Analysis of the far infrared spectrum[15] and a microwave study[16] both agree on a barrier of 0.75 kcal mol⁻¹ for the ring inversion of trimethylene sulphide (thietane) (5). In contrast to what is observed for oxetane (1) the first four vibrational levels of thietane fall below the barrier.

Several groups have worked on the conformational analysis of substituted thietanes. Dodson *et al.* prepared *cis*- and *trans*-2,4-diphenyl thietane (6 and 7), their 1-oxides (8 and 9) and 1,1-dioxides (10 and 11).[17] Base-catalysed equilibration of the oxides gave 96% or more of (8) and (10). On the basis of this evidence and after analysis of some ^{1}H NMR spectra, it was deduced that both the phenyl and S=O groups have a substantial equatorial preference in these compounds.

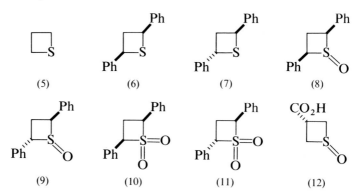

X-Ray diffraction has been employed in a structural study of *trans*-3-carboxythietane-1-oxide (12).[18] In the solid phase the carboxyl group is axial and the S=O is equatorial. This equatorial preference of the sulphoxide group was also found by Siegl and Johnson[19] in other studies of thietane-1-oxides.*

Trimethylene Selenide

A barrier to ring puckering of 1.08 kcal mol^{-1} in trimethylene selenide (13) was deduced from infrared and Raman spectroscopic studies.[20]

$$\boxed{}\text{—Se}$$
(13)

Silacyclobutane

Microwave spectra and electron diffraction results show a non-planar silacyclobutane ring (14) with a puckering angle of 29°.[21,22] Infrared spectra[23] agree with the microwave results[21] in assigning a barrier to inversion of 1.26 kcal mol^{-1}, but give a slightly different puckering angle of

---

* The inference in Moriarty's otherwise excellent review that this work showed an axial preference is incorrect.[1]

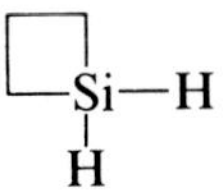

$$\overset{\displaystyle \boxed{}}{\underset{\displaystyle H}{-Si-H}}$$

(14)

$37 \pm 2°$. Electron diffraction studies on some substituted derivatives have been reported.[24]

Azetidine

Azetidine (15) is an interesting example of a molecule that inhabits a potential-energy well with two unequal energy minima and that obeys eq. 3.3. The far-infrared spectra obtained by Carriera and Lord,[25] were interpreted in terms of two minima differing in energy by 0.27 kcal mol^{-1}

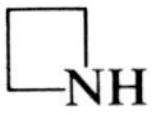

$$\boxed{}-NH$$

(15)

and separated by a barrier (ground state → transition state) of 1.26 kcal mol^{-1}. It is presumed, but not established,* that the lower-energy form has N—H equatorial. The structures of several salts based on the azetidine ring have been determined by X-ray crystallography.[26]

Phosphetane

There appears to have been no work published on the parent compound; however, some studies of substituted derivatives have appeared.[27-29] The rates of pyramidal inversion at phosphorous have been measured by isomerization experiments in some 2,2,3,4,4-pentamethyl derivatives (16).[27] When R = methyl the rate of inversion is so slow as to be undetectable. However when R is phenyl or t-butyl the rate of inversion is measurable. This is attributed to a reduction in the activation energy by either conjugation in the transition state lowering its energy R = Ph, or steric destabilization of the ground state raising its energy, R = t-Bu.

The structures of some phosphetane-1-oxides[28] and oxaphosphetanes[29] have been determined crystallographically.

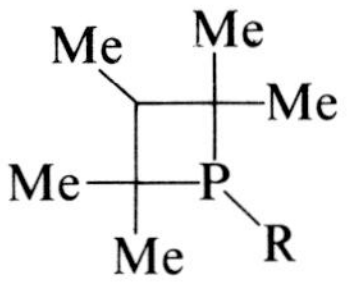

$$\begin{matrix} Me & Me & \\ & | & \\ Me- & \boxed{} & -Me \\ & | & \\ Me & R & \end{matrix}$$

(16)

* See, however, J. Catalan, O. Mo and M. Yaney. (1978). *J. Molec. Struct.* **43**, 251.

Oxazetidines

Three groups have reported measurements of nitrogen inversion barriers in perfluorooxazetidines (17).[30–32] When the N-substituent is alkyl or fluoroalkyl rates of nitrogen inversion are measurable by NMR spectroscopy. When the N-substituent is a halogen the rate of nitrogen inversion is considerably reduced and there is evidence of two conformations (N—X axial and N—X equatorial) being present.

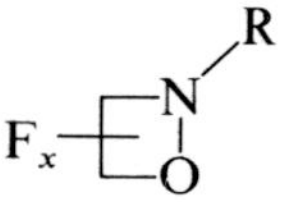

$$F_x \overline{\underset{\underset{O}{|}}{\overset{\overset{N}{|}}{}}}^{\displaystyle R}$$

$$x = 1 - 4$$

(17)

1,3-Dithietane

1,3-Dithietane (18) and all its oxidation products on sulphur, including the *cis*- and *trans*-1,3-dioxides, have been prepared and investigated.[33] Interestingly no microwave absorption spectrum is observed for (18). Since a non-planar ring would possess a reasonable dipole moment this suggests that 1,3-dithietane is a planar molecule reminiscent of 3,3-difluorooxetane (2).[11] X-Ray diffraction shows the mono-oxide (19) to have a non-planar structure with equatorial S=O and a puckering angle of 39°, whereas the tetraoxide (20) is also revealed to be planar.

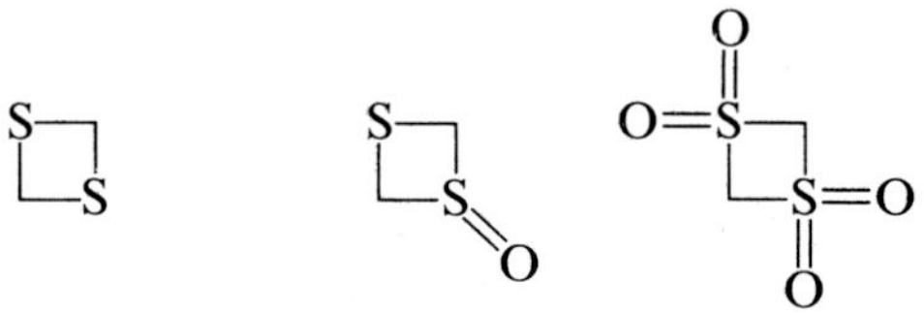

(18)   (19)   (20)

Five-Membered Rings

Introduction

Whereas for four-membered rings only one coordinate is needed to describe the ring puckering vibration, two coordinates are required for an adequate conformational description of five-membered rings.[34]

If cyclopentane were a planar molecule, the internal C—C—C angles of the ring would be those of the regular pentagon (108°) and would differ so little from those of a regular tetrahedron (109.5°) or "strain-free" open chain hydrocarbons (112.7°) that bond angle deformation should contribute little to the overall strain energy. The planar molecule would, however, possess considerable strain arising from the five perfectly eclipsed C—C bonds. There are two ways that cyclopentane can deform to relieve this strain whilst still retaining some of its original symmetry. Displacement of one carbon atom above or below the plane of the other four gives the envelope form (21) with a plane of symmetry, C_s. Alternatively, displacement of two adjacent carbon atoms equal distances on either side of the plane of the remaining three gives the half-chair form (22) with a two-fold axis of symmetry, C_2.

(21)   (22)

In 1947 Kilpatrick, Pitzer, and Spitzer[35] showed that the puckering of the ring is not of a definite type, i.e. neither (21) nor (22) is more stable, but that the angle of maximum puckering rotates around the ring. Thus the two coordinates required to specify the conformation of a five-membered ring can be taken as two phase angles which measure (i) the position of maximum puckering in the ring, and (ii) the amplitude of the puckering. The second angle really represents what is just a normal vibration of the ring, whilst the first angle specifies its position.

Calculations[36] indicate that the envelope and half-chair forms are of very similar energy, and that there is little energy barrier between them. The movement of the molecule from the C_s to C_2 forms as the first phase angle varies is called a pseudorotation. In a pseudorotation the atoms themselves do not rotate, it is the phase of the puckering that rotates around the ring. Pseudorotation has been defined by Hendrickson as the passage of a ring with a plane of symmetry through an atom, to one with an axis of symmetry bisecting the bond adjacent to the atom and vice versa.[36b] A more general definition of ring puckering coordinates has been given by Cremer and Pople,[37] who then used this theory to perform molecular orbital calculations for the pseudorotation of five-membered rings. Pseudorotation in cyclopentane has been confirmed by many subsequent workers.[38]

In the conformational analysis of heterocyclic five-membered rings we are concerned with trying to establish the way that the potential energy of the molecule varies as it pseudorotates, with finding the lowest-energy conformation(s) on the pseudorotation circuit, and with determining the influence these factors have on the chemistry of the molecules.

Oxygen-containing rings

The fundamental physical evidence on tetrahydrofuran (THF) (23) shows that pseudorotation is essentially free. Greenhouse and Strauss[39] studied the far infrared spectra of THF and 1,3-dioxolan (24) in the region 20–370 cm^{-1}. Both were found to have barriers of ca 0.1 kcal mol^{-1} confirming earlier interpretations.[40] Similarly, electron diffraction on THF shows that: "the molecule exhibits nearly free pseudorotation, i.e. that the energy difference between C_s and C_2 is small (say 0.0 ± 0.3 kcal mol^{-1})."[41]

(23) 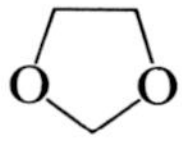   (24)

Halogenotetrahydrofurans, on the other hand, show evidence of greater restriction to pseudorotation with energy wells that are steeper than those of the corresponding halogenocyclopentanes. This means that the amplitudes of the molecular vibrations in the energy wells, known as pseudo-librations, will be smaller.[42]

Arguably the most important series of oxygen-containing five-membered ring compounds is the furanoid sugars. Several reviews have appeared covering the conformational analysis of these compounds.[43] X-Ray crystallographers have devoted some attention to studying the nucleosides.[44–48] In particular cytidine and the cytidylic acids have been examined in some detail. The D-ribose ring in cytidine (25) has C(3) 50 pm out of the plane of the other four ring atoms.[45] In the cytidylic acids (26) two different ribose ring envelope conformations are observed,[46] one with C(3) and the other with C(2)

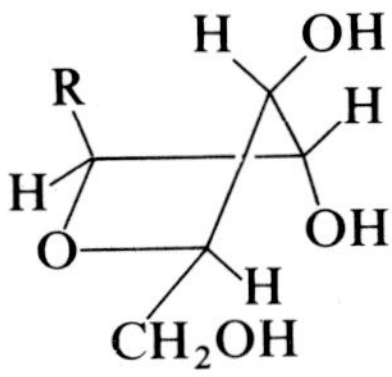

(25) 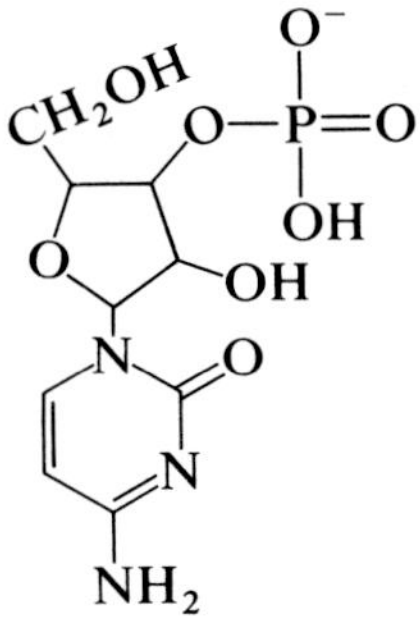   (26)

ıt of the plane of the other four atoms. Cyclization to the cytidine cyclic ıosphate (27) gives a bicyclic ring conformation in which the oxygen atom out of the plane of the other four.[47] A Karplus type of equation has been :veloped for the furanoid ring in nucleosides and nucleotides which allows ł–¹H vicinal coupling constants to be interpreted in terms of the dihedral ıgles in the carbohydrate ring.[48]

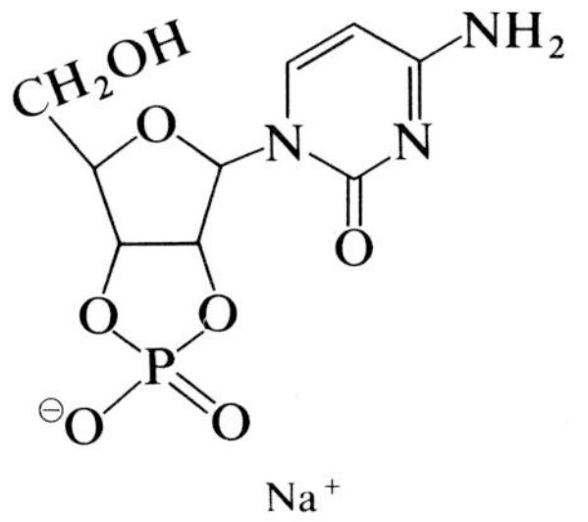

(27) 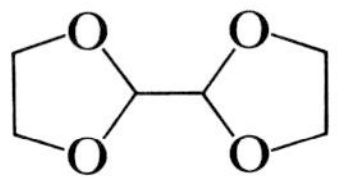   (28)

1,3-Dioxolan has been shown to pseudorotate essentially freely in the ıpour phase.[39] 2,2'-Bis-1,3-Dioxolan (28) is shown by X-ray crystallo-ʾaphy to have a conformation in the solid mid-way between the half-chair ıd envelope forms.[49] On the other hand, the related compound 2-oxo-1,3-ɩoxolan (29) shows a half-chair conformation in the solid with each ethylene carbon ca 30 pm out of the O—C—O plane.[50] This result is ɔnfirmed by microwave spectroscopy[51] for molecules in the vapour phase, ıd by ¹³C chemical shift and coupling constant measurements for mole-ıles in solution.[52] Analysis of the AA'BB' NMR spectra of the ring hydro-ɪn atoms in some 1,3-dioxolan derivatives is in agreement with a puckered ɩng.[53,54]

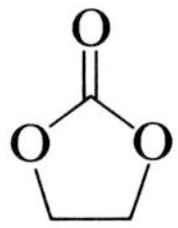

(29) 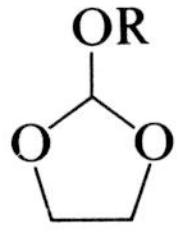   (30)

In some 2-alkoxy-1,3-dioxolans (30) Altona has shown from a combina-ɔn of NMR and dipole moment data that there are *anti* and *gauche* forms ɔout the exocyclic C(2)—O bond.[54] The torsion angle about the ring C—C ɔnd decreases on going from an *anti* to a *gauche* C(2)—O conformation. ɪltona pointed out two conflicting conformational tendencies in 1,3-di-ҡolan derivatives: a tendency to increase the CH_2—CH_2 torsion angle, ; this bond has the greatest barrier to internal rotation of those in the ng, and a tendency of the O—C—O fragment to minimize electrostatic ɩteractions by assuming a chiral conformation. Both ends of the ring are ıus competing for the available amount of puckering.[54]

The acetal portion of the 1,3-dioxolan ring renders derivatives amenab]
to acid-catalysed equilibration; Eliel[55] and Anteunis[56] have used th
technique. In general, free energy differences between *cis* and *trans* isome**I**
are low, being 0.3–0.5 kcal mol^{-1} for dialkyl, and 0.2–0.7 kcal mol^{-1} fc
trialkyl derivatives. The data are consistent with a highly flexible fiv**e**
membered ring, with only the most bulky substituents showing signs **c**
specific interactions.

Microwave spectra on the ozonides of ethylene,[57] propylene and *trans-2*
butene[58] show that the 1,2,4-trioxolan ring prefers a half-chair conformatio
(31) with the C—O—C portion of the ring forming the reference plane, an
that alkyl substituents prefer the equatorial position.

(31)   (32)

Although in most derivatives of butyrolactone the lactone group **I**
almost essentially planar, in certain derivatives such as glucuronolacton**e**
(32) constraints from steric factors in the ring can cause deviations of atom**i**
positions from the best fit plane through the lactone group of up to 10 pm.[5]
The conformation of the lactone ring in (32) is, however, best described as a
envelope with the flap atom (C(3)) 62 pm out of the best fit plane of th**e**
lactone group.[59] This view of the butyrolactone ring has been confirmed b
NMR studies on a wide variety of other derivatives.[60]

Sulphur-containing rings

Thiophane (33) and 1,3-dithiolan (34) show evidence of a much greate**r**
restriction of pseudorotation than do their oxygen-containing counterpart
discussed above. Thermodynamic studies of thiophane suggest a barrier t
pseudorotation of ca 2.8 kcal mol^{-1}. The contribution that this barrie
makes to the thermodynamic properties of the compound is essentiall**y**
identical to that of a normal restricted rotation.[61] This view was late
confirmed by spectroscopic data.[62] Information as to which conformatio**n**
is preferred was provided by Seip's electron diffraction data and molecula**r**

(33)   (34)

mechanics calculations.[63] These methods both show the half-chair (C_2) form to be 2–3 kcal mol^{-1} more stable than the envelope (C_s) form. An X-ray diffraction study of the 1:1 thiophane–bromine complex shows a seemingly planar ring.[64] This is attributed to the superposition in the unit cell of images from two static or dynamically interconverting C_2 ring conformations. This explanation is supported by the large anisotropic temperature factors observed for the C(2) and C(5) atoms.

1,3-Dithiolan (34) derivatives have been studied by X-ray crystallography[65] and by vibrational spectroscopy.[66] In the region 900–300 cm^{-1} several commonly occurring bands were found from which it was concluded that the molecules possess a non-planar skeleton. Although no firm conclusions were possible as to the basic conformation of the ring, it was thought that the most important conformation probably is of symmetry C_2 (half-chair). Studies on the acid-catalysed equilibration of some 2-alkyl-4-methyl and 2-alkyl-2,4-dimethyl-1,3-dithiolans have been reported.[67] The data obtained, supported by some NMR work, suggest that the dithiolane ring may be quite flexible and that a minimum energy conformation is only well defined if there is a bulky substituent at the 2-position.

There is strong evidence of 1,3-oxathiolan (35) derivatives having preferred conformations, which are probably best described as envelopes. X-Ray diffraction on the spirosteroid (36) shows an envelope conformation of the oxathiolane ring with the CH_2 adjacent to oxygen 50 pm outside the plane defined by the other four atoms.[68] Nuclear magnetic resonance spectral analysis of some 2- and 2,2-substituted derivatives is in accord with rapid pseudorotation and two envelope conformations (37) and (38) being preferred.[69] These conclusions are supported by Pihlaja's results from chemical equilibrations and by spectroscopic studies in this series.[70]

X-Ray crystallographic studies on ethylene sulphate (39) show a molecule

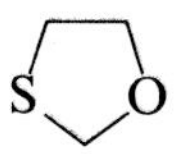

(35)

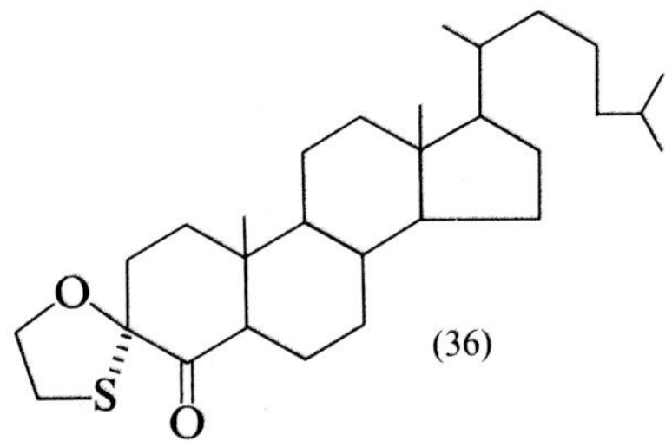

(36)

(37)

(38)

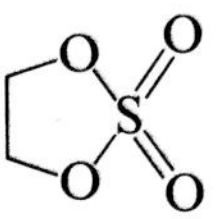

(39)

with C_2 symmetry. The C—C bond is at an angle of 20.6° to the plane of tl O—S—O atoms in the ring.[71]

Nitrogen-containing rings

A thermodynamic study of pyrrolidine (40) reveals a barrier to pseud rotation of ca 0.3 kcal mol^{-1}.[72] Nuclear magnetic resonance studies cis- and trans-4-hydroxy-L-proline (41) suggest that the molecules of tl trans isomer exist in envelope conformations with C(4) as the flap atom ar the hydroxyl group axial.[73] The cis isomer, which probably has a differei conformation, also has an axial hydroxyl group. A careful NMR investig tion of nicotine (42) suggests an envelope conformation of the pyrrolidir ring with the N-methyl and pyridyl groups in equatorial positions.[74]

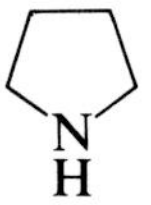

(40) 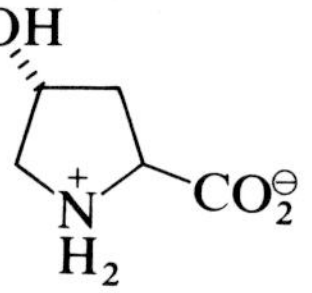   (41) 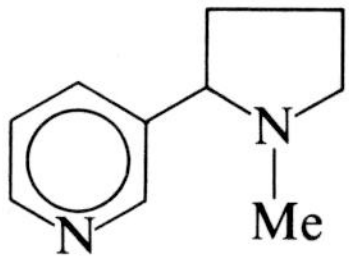   (42)

Nitrogen inversion rates in five-membered rings have attracted coi siderable attention.[75]

Other nitrogen-containing five-membered rings that have been studie include 1,2-oxazolidines,[76] 1,3-oxazolidines,[77] 1,3-thiazolidines,[78] pyr zolidines,[79] pyrrolizidines,[80] and 1,3,4-oxadiazolidines.[81]

In the biologically important field of peptides incorporating a pyrrolidir ring in a proline residue, NMR measurements show that the conformatio of the pyrrolidine ring is a time average of a series of puckered conformatioi which are highly dependent on the presence and nature of the adjoinir residues.[82]

Phosphorus-containing rings

Much of our evidence on the conformations of five-membered phosphoru containing rings comes from NMR spectral data on phospholanes wi1 further heteroatoms surrounding the phosphorus, although a report h; appeared showing clear evidence of the conformational dependence of [31] chemical shifts in some alkylated derivatives of phospholane.[83]

Gas phase electron diffraction studies are consistent with an envelo conformation having P at the flap for 2-chloro-1,3,2-dioxaphospholan (43)[84] as well as for the closely related 1,3-dithiophospholane (44)[85] an N-methyl-1,3,2-oxazaphospholane (45).[86] However 2-chloro-4,4,5,5-tetr;

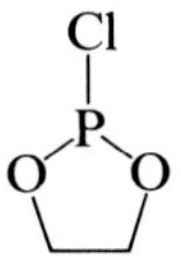

(43) 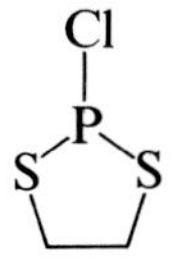 (44) 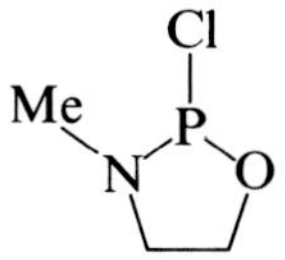 (45)

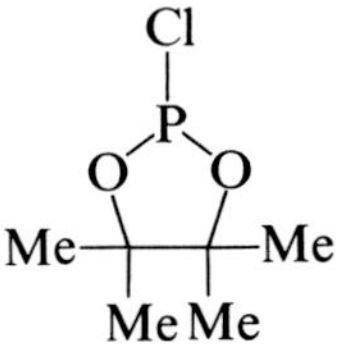

(46) 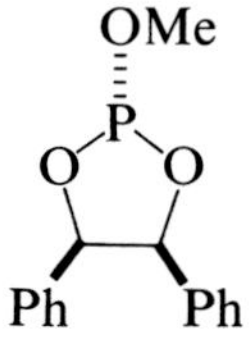 (47)

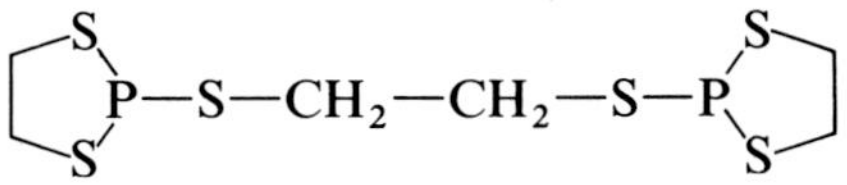

$$\text{S}\diagdown\text{P}-\text{S}-\text{CH}_2-\text{CH}_2-\text{S}-\text{P}$$

(48)

methyl-1,3,2-dioxaphospholane (46) appears to be in a half-chair conformation.[87]

X-Ray crystallography has established an envelope C-flap conformation for 4,5-diphenyl-*trans*-2-methoxy-1,3,2-dioxaphosphorinane (47),[88] whilst for 2,2′-(ethylenedithio)*bis*-1,3-dithio-2-phospholane (48) the structure is close to that of an S flap envelope.[89]

Nuclear magnetic resonance spectra have given evidence on phospholanes with oxygen, nitrogen, and sulphur atoms surrounding the phosphorous.[90-101]

Heterocyclic derivatives of cyclopentene

In five-membered rings containing a double bond only one coordinate is needed to specify the ring conformation. This is because the planar nature of the double bond raises the frequency of vibrations associated with one of the two coordinates associated with a five-membered ring so much that it can be neglected for conformational purposes. Cyclopentene has been shown to be a non-planar molecule with a dihedral angle of ca 22° and a barrier to inversion of ca 0.66 kcal mol^{-1}.[102,103] Since this barrier presumably arises from torsional strain it is expected to be modified in heterocyclic systems as the barriers to rotation about carbon–heteroatom bonds change. Thus 2,3-dihydrofuran (49) shows a barrier of 0.24 kcal mol^{-1} and a dihedral angle of 19° [103,104] whilst 2,5-dihydrofuran (50) appears to be a planar molecule.[103] Several other heterocyclic derivatives of cyclopentene that are planar have been found.[72,105]

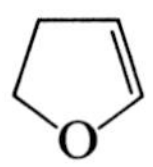

(49)  (50)

References

1a. R. M. Moriarty. (1974). *In* "Topics in Stereochemistry", Vol. 8, (Eds E. L. Eliel and N. L. Allinger), 271. Wiley Interscience, New York.

1b. T. B. Malloy, L. B. Baumann and L. A. Carreira. (1979). *In* "Topics in Stereochemistry," Vol. 11, (Eds. E. L. Eliel and N. L. Allinger), 97. Wiley Interscience, New York.

2. A. von Baeyer. (1885). *Chem. Ber.* **18**, 2277.

3. H. Sachse. (1890). *Chem. Ber.* **23**, 1363; (1892). *Z. Phys. Chem.* **10**, 203.

4. T. P. Wilson. (1943). *J. Chem. Phys.* **11**, 369.

5. R. P. Bell, personal communication.

6. R. P. Bell. (1945). *Proc. R. Soc.* **A183**, 328.

7. W. F. Edgell and D. G. Weiblen. (1950). *J. Chem. Phys.* **18**, 511.

8. J. D. Dunitz and V. Schomaker. (1952). *J. Chem. Phys.* **20**, 1703; F. A. Miller and R. J. Capwell. (1971). *Spectrochim. Acta.* **27A**, 947; F. A. Miller, R. J. Capwell, R. C. Lord and D. G. Rea. (1972). *Spectrochim. Acta* **28A**, 603.

9. S. Meiboom and L. C. Snyder. (1967). *J. Am. Chem. Soc.* **89**, 1038; (1970). *J. Chem. Phys.* **52**, 3857.

10. W. D. Gwinn, J. Zinn and J. Fernandez. (1959). *Bull. Am. Phys. Soc.* **4**, 153; S. I. Chan, J. Zinn and W. D. Gwinn. (1960). *J. Chem. Phys.* **33**, 295; S. I. Chan, J. Zinn, J. Fernandez and W. D. Gwinn. (1960). *J. Chem. Phys.* **33**, 1643.

11. G. L. McKown and R. A. Beaudet. (1971). *J. Chem. Phys.* **55**, 3105.

12. D. R. Lide, Jr. (1959). *J. Chem. Phys.* **30**, 37.

13. J. Jokisaari, E. Rohkaama and P. O. I. Virlanen. (1970). *Suomen Kem. B* **43**, 219.

14. K. Pihlaja, K. Polviander, R. Keskinen and J. Jalonen. (1971). *Acta Chem. Scand.* **25**, 765.

15. T. R. Borgers and H. L. Strauss. (1966). *J. Chem. Phys.* **45**, 947.

16. D. O. Harris, H. W. Harrington, A. C. Luntz and W. D. Gwinn. (1966). *J. Chem. Phys.* **44**, 3467.

17. R. M. Dodson, E. H. Jancis and G. Klose. (1970). *J. Org. Chem.* **35**, 2520.

18. S. Allenmark. (1966). *Ark. Kemi* **26**, 73; S. Abrahamson and G. Rehnberg. (1972). *Acta Chem. Scand.* **26**, 494.

19. W. O. Siegl and C. R. Johnson. (1971). *Tetrahedron* **27**, 341.

20. A. B. Harvey, J. R. Durig and A. C. Morrissey. (1967). *J. Chem. Phys.* **47**, 4864; (1969). **50**, 4949.

21. W. C. Pringle, Jr. (1971). *J. Chem. Phys.* **54**, 4979.

22. L. V. Vilkov, V. S. Mastryukov, Tu. V. Baurova and V. M. Vdovin. (1967). *Dokl. Akad. Nauk. SSSR* **177**, 1084 (Engl. translation p. 1146).

23. J. Laane and R. C. Lord. (1968). *J. Chem. Phys.* **48**, 1508.

24. L. V. Vilkov, V. S. Mastryukov, V. D. Oppenheim and N. A. Tarasenko, unpublished work quoted in reference 1.

25. L. A. Carreira and R. C. Lord. (1969). *J. Chem. Phys.* **51**, 2735.

26. H. M. Berman, E. L. McGandy, J.W. Burgner, II, and R. L. van Etten. (1969). *J. Am. Chem. Soc.* **91**, 6173, 6177; R. L. Townes and L. M. Tefonas. (1971). *J. Am. Chem. Soc.* **93**, 1761; R. L. Snyder, E. L. McGandy, J. W. Burgner, II and R. L. van Etten. (1969). *J. Am. Chem. Soc.* **91**, 6187; H. M. Zacharis and L. M. Tregonas. (1971). *J. Am. Chem. Soc.* **93**, 2935.

27. S. E. Cremer, R. J. Chorvat, C. H. Chang and D. W. Davis. (1968). *Tetrahedron Lett.* 5799.

28.   M. U-Haque. (1970). *J. Chem. Soc. B* 938; (1971). 117
29.   M. U-Haque, C. N. Caughlan, F. Ramirez, J. F. Pilot and C. P. Smith. (1971). *J. Am. Chem. Soc.* **93**, 5229.
30.   S. Andreades. (1962). *J. Org. Chem.* **27**, 4163.
31.   J. Lee and K. G. Orrell. (1965). *Trans. Faraday Soc.* **61**, 2342.
32.   J. D. Readio and R. A. Falk. (1970). *J. Org. Chem.* **35**, 927; J. D. Readio. (1970). *J. Org. Chem.* **35**, 1607.
33.   E. Block, E. R. Corey, R. E. Penn, T. L. Renken and P. F. Sherwin. (1976). *J. Am. Chem. Soc.* **98**, 5715.
34.   For a review see B. Fuchs. (1978). *In* "Topics in Stereochemistry," Vol. 10, (Eds E. L. Eliel and N. L. Allinger), 1. Wiley Interscience, New York.
35.   J. E. Kilpatrick, K. S. Pitzer and R. Spitzer. (1947). *J. Am. Chem. Soc.* **69**, 2483.
36a. H. J. Geise, C. Altona and C. Romers. (1967). *Tetrahedron Lett.* 1383; S. Lifson and A. Washell. (1968). *J. Chem. Phys.* 49, 5116; J. B. Hendrickson. (1961). *J. Am. Chem. Soc.* **83**, 4537; (1963). **85**, 4059.
36b. J. B. Hendrickson. (1967). *J. Am. Chem. Soc.* **89**, 7047.
37.   D. Cramer and J. A. Pople. (1975). *J. Am. Chem. Soc.* **97**, 1354, 1358.
38.   J. P. McCullough. (1958). *J. Chem. Phys.* **29**, 966; J. P. McCullough, R. E. Pennington, J. C. Smith, I. A. Hossenlop and G. Waddington. (1959). *J. Am. Chem. Soc.* **81**, 5880; W. J. Adams, H. J. Geisse and L. S. Bartell. (1970). *J. Am. Chem. Soc.* **92**, 5013; J. R. Durig and D. W. Wertz. (1968). *J. Chem. Phys.* **49**, 5116.
39.   J. A. Greenhouse and H. L. Strauss. (1969). *J. Chem. Phys.* **50**, 124.
40.   W. J. Lafferty, D. W. Robinson, R. V. St. Louis, J. W. Russell and H. L. Strauss. (1965). *J. Chem. Phys.* **42**, 2915.
41.   A. Almenningen, H. M. Seip and T. Willadsen. (1969). *Acta Chem. Scand.* **23**, 2748.
42.   H. R. Buys, C. Altona and E. Havinga. (1968). *Tetrahedron* **24**, 3019.
43.   L. D. Hall. (1964). *Adv. Carbohyd. Chem.* **19**, 51; G. A. Jeffrey and R. D. Rosenstein. (1964). *Adv. Carbohyd. Chem.* **19**, 7; J. W. Green. (1966). *Adv. Carbohyd. Chem.* **21**, 95; M. Sundaralingham. (1965). *J. Am. Chem. Soc.* **87**, 599; P. Murray-Rust and S. Motherwell. (1978). *Acta Cryst,* **B34**, 2534.
44.   S. Furberg. (1950). *Acta Chem. Scand.* **4**, 751.
45.   S. Furberg. (1950). *Acta Cryst.* **3**, 325.
46.   E. Alver and S. Furberg. (1959). *Acta Chem. Scand.* **13**, 910; S. Furberg. (1960). *Acta Chem. Scand.* **14**, 189.
47.   C. L. Coulter. (1973). *J. Am. Chem. Soc.* **95**, 570.
48.   C. Altona and M. Sundaralingham. (1973). *J. Am. Chem. Soc.* **95**, 2333.
49.   S. Furberg and O. Hassel. (1950). *Acta Chem. Scand.* **4**, 1584.
50.   C. J. Brown. (1954). *Acta Cryst.* **7**, 92.
51.   I. Wang, C. O. Britt and J. E. Boggs. (1965). *J. Am. Chem. Soc.* **87**, 4950.
52.   K. Pihlaja and K. Rossi. (1977). *Acta Chem. Scand.* **B31**, 899.
53.   B. Matthiason. (1963). *Acta Chem. Scand.* **17**, 2133; D. Gagnaire and J. B. Robert. (1965). *Bull. Soc. Chim. France* 3646; R. J. Abraham. (1965). *J. Chem. Soc.* 256.
54.   C. Altona and A. P. M. van der Veek. (1968). *Tetrahedron* **24**, 4377.
55.   E. L. Eliel and W. E. Willy. (1969). *Tetrahedron Lett.* 1775; W. E. Willy, G. Binsch and E. L. Eliel. (1970). *J. Am. Chem. Soc.* **92**, 5394.
56.   Y. Rommelaere and M. Anteunis. (1970). *Bull. Soc. Chim. Belges* **79**, 11; G. Lemiere and M. Anteunis. (1971). *Bull. Soc. Chim. Belges* **80**, 215.
57.   C. W. Gillies and R. L. Kuckowski. (1972). *J. Am. Chem. Soc.* **94**, 6337, 7609.

58. R. P. Lattimer, R. L. Kuckowski and C. W. Gillies. (1974). *J. Am. Chem. Soc.* **96**, 348.
59. G. A. Jeffrey and S. H. Kim. (1966). *Chem. Comm.* 211.
60. R. N. Johnson, J. B. Lowry and N. V. Riggs. (1967). *Tetrahedron Lett.* 5113; K. Ohga and T. Matsuo. (1973). *Bull. Chem. Soc. Jpn.* **46**, 2181; K. Pihlaja, F. G. Riddell, J. Jalonen, P. Rinta-Panttila and M. Anteunis. (1974). *Org. Mag. Res.* **6**, 301.
61. W. N. Hubbard, H. L. Finke, D. W. Scott, J. P. McCullough, C. Katz, M. E. Gross, J. F. Messerly, R. E. Pennington and G. Waddington. (1952). *J. Am. Chem. Soc.* **74**, 6025.
62. G. A. Crowder and D. W. Scott. (1965). *J. Molec. Spectrosc.* **16**, 122.
63. Z. Nahlovska, B. Nahlovsky and H. M. Seip. *Acta Chem. Scand.* (1969). **23**, 3534; (1970). **24**, 1903.
64. G. Allegra, G. E. Wilson, Jr, E. Benedetti, C. Pedone and R. Albert. (1970). *J. Am. Chem. Soc.* **92**, 4002.
65. L. B. Brahde. (1954). *Acta Chem. Scand.* **8**, 1145.
66. H. Fuhrer and H. H. Gunthard. (1962). *Helv. Chim. Acta.* **45**, 2036.
67. R. Keskinen, A. Nikkila and K. Pihlaja. (1973). *J.C.S. Perkin II* 1376.
68. A. Cooper and D. A. Norton. (1968). *J. Org. Chem.* **33**, 3535.
69. G. E. Wilson, Jr, M. G. Huang and F. A. Bovey. (1970). *J. Am. Chem. Soc.* **92**, 5907.
70. R. Keskinen, A. Nikkila and K. Pihlaja. (1972). *Tetrahedron* **28**, 3943; (1977). *J.C.S. Perkin II* 343; K. Pihlaja. (1970). *Suomen Kemistilehti* **43B**, 143; K. Pihlaja, T. Nurmi and P. Pasanen. (1977). *Acta Chem. Scand.* **B31**, 895.
71. F. P. Boer, J. J. Flynn, E. T. Kaiser, O. R. Zaborsky, D. A. Thomalin, A. E. Young and Y. C. Tong. (1968). *J. Am. Chem. Soc.* **90**, 2970.
72. J. P. McCullough, D. R. Douslin, W. N. Hubbard, S. S. Todd, J. F. Messerly, I. A. Hossenlopp, F. R. Frow, J. P. Dawson and G. Waddington. (1959). *J. Am. Chem. Soc.* **81**, 5884.
73. R. J. Abraham and K. A. McLauchlan. (1962). *Molec. Phys.* **5**, 195.
74a. T. P. Pitner, W. B. Edwards, III, R. L. Bassfield and J. F. Whidby. (1978). *J. Am. Chem. Soc.* **100**, 246.
74b. J. F. Whidby, W. B. Edwards and T. P. Pitner. (1979). *J. Org. Chem.* **44**, 794.
75. J. M. Lehn and J. Wagner. (1970). *Tetrahedron* **26**, 4227; K. B. Nielsen (1973). *Acta Chem. Scand.* **27**, 1092; J. M. Lehn. (1970). *Fortschritte* **15**, 311; V. J. Baker, A. R. Katritzky, J.-P. Majoral, S. F. Nelsen and P. J. Hintz. (1974). *J.C.S. Chem. Comm.* 823; F. G. Riddell, J. M. Lehn and J. Wagner. (1968). *Chem. Comm.* 1403; J. B. Lambert, W. L. Oliver and B. S. Packard. (1971). *J. Am. Chem. Soc.* **93**, 933.
76. V. M. Shitkin, S. L. Joffe, M. V. Kashutna and V. A. Tartakovskii. (1977). *Izv. Akad. Nauk. SSSR, Ser Khim* 2266; *Chem. Abs.* **88**, 50151n.
77. S. L. Spassov, J. N. Stefanovsky, B. J. Kurtev and G. Fodor. (1972). *Chem. Ber.* **105**, 2462.
78. G. E. Wilson, Jr. and T. J. Bazzone. (1974). *J. Am. Chem. Soc.* **96**, 1465; V. M. Kulkarni and G. Govil. (1977). *J. Pharm. Sci.* **66**, 483.
79. S. Caccamese, P. Finocchiaro, P. Maravigna and G. Montando. (1977). *Gaz. Chim. Ital.* **107**, 415.
80. A. J. Aasen, C. C. J. Culvenor and R. I. Willing. (1971). *Aust. J. Chem.* **24**, 2575.
81. V. J. Baker, A. R. Katritzky and J. D. Majoral. (1975). *J.C.S. Perkin II* 1191.
82. R. Deslauriers, I. C. P. Smith and R. Walter. (1974). *J. Biol. Chem.* **249**, 7006.
83. J. J. Breen, J. F. Engel, D. K. Myers and L. D. Quin. (1972). *Phosphorus* **2**, 55.

84. V. A. Naumov, N. M. Zaripov and V. G. Dashevskii. (1969). *Dokl. Akad. Nauk SSSR* **188**, 1062.
85. G. Y. Schultz, I. Hargittai, J. Martin and J. B. Robert. (1974). *Tetrahedron* **30**, 2365.
86. V. A. Naumov and M. A. Pudovik. (1972). *Dokl. Akad. Nauk SSSR* **203**, 351.
87. V. A. Naumov and N. M. Zaripov. (1970). *Z. Struct. Khim.* **11**, 1124.
88. M. G. Newton and B. S. Campbell. (1974). *J. Am. Chem. Soc.* **96**, 7790.
89. M. G. Newton, H. C. Brown, C. J. Finder, J. B. Robert and J. Martin. (1974). *J. Chem. Soc. Chem. Comm.* 455.
90. K. Bergesen and T. Vikane. (1972). *Acta Chem. Scand.* **26**, 2153.
91. R. H. Cox, B. S. Campbell and M. G. Newton. (1972). *J. Org. Chem.* **37**, 1557.
92. P. Haake, J. P. McNeal and E. J. Goldsmith. (1968). *J, Am. Chem. Soc.* **90**, 715.
93. D. Gagnaire, J. B. Robert, J. Verrier and R. Wolf. (1966). *Bull. Soc. Chim. France* 3719.
94. K. Bergesen, M. Bjoroy and T. Gramstad. (1972). *Acta Chem. Scand.* **26**, 3037.
95. S. C. Peake, M. Fild, R. Schmutzler, R. K. Harris and R. G. Rees. (1972). *J.C.S. Perkin II* 380.
96. W. G. Bentrude and H. W. Tan. (1975). *Tetrahedron Lett.* 619.
97. W. G. Bentrude and H. W. Tan. (1976). *J. Am. Chem. Soc.* **98**, 1850.
98. J. P. Albrand, A. Cogne, D. Gagnaire, J. Martin, J. B. Robert and J. Verrier. (1971). *Org. Mag. Res.* **3**, 75.
99. J. P. Albrand, D. Gagnaire, J. Martin and J. B. Robert. (1973). *Org. Mag. Res.* **5**, 33.
100. J. H. Hargis, S. P. Worley, W. B. Jennings and M. S. Tolley. (1977). *J. Am. Chem. Soc.* **99**, 8090.
101. G. A. Gray and T. A. Albright. (1977). *J. Am. Chem. Soc.* **99**, 3243.
102. C. W. Rathjens, Jr. *J. Chem. Phys.* (1962). **36**, 2401; L. H. Scharpen. *J. Chem. Phys.* (1968). **48**, 3552; J. Laane and R. C. Lord. *J. Chem. Phys.* (1967). **47**, 4941.
103. T. Ueda and T. Shimanouchi. (1967). *J. Chem. Phys.* **47**, 5018.
104. W. H. Green. (1969). *J. Chem. Phys.* **50**, 1619.
105. J. Laane. (1965). *J. Chem. Phys.* **50**, 776; (1970). **52**, 358; W. H. Green and A. B. Harvey. (1968). *J. Chem. Phys.* **49**, 177; K. L. Dorris, C. O. Britt and J. E. Boggs. (1966). *J. Chem. Phys.* **44**, 1352.

4

Six-Membered
Oxygen-Containing
Rings

General Considerations

Six-membered saturated oxygen-containing rings occur quite widely in nature, the most abundant type being the pyranose sugars. One of the reasons why the conformational analysis of model oxygen-containing rings has been so widely investigated is to understand better in simpler systems, what goes on in the more complicated carbohydrates.

Compared with carbocyclic systems the oxygen-containing rings have shorter C—O bonds and smaller oxygen atoms. As pointed out in Chapter 2, these features will increase certain non-bonded repulsions and reduce others. Other effects mentioned in Chapter 2 that are brought out clearly in oxygen-containing rings are those arising from hydrogen-bonding and dipolar interactions.

If one formally regards ether oxygen as being sp^3 hybridized then there are two pairs of electrons on oxygen in a six-membered ring pointing in axial and equatorial directions. This may or may not be a good description and the matter has been debated in the literature.[1,2] To the current author it seems that for non-bonded interactions and certain bond rotation processes a single electron cloud incorporating both pairs of electrons is an adequate description, but that in much stronger interactions involving chemical bonding to the oxygen atoms, two directed pairs of electrons are an essential description.

The conformational consequences of replacing ring methylene groups by oxygen atoms have been reviewed by Dale.[3] The picture presented by Dale is essentially qualitative. A more quantitative approach has been attempted by Allinger et al.[4] who have developed a molecular mechanics force-field that gives reasonable predictions of some aspects of the conformational analysis of oxygen-containing rings.

With oxygen atoms situated 1,3 in a carbon chain or in a ring the anomeric effect comes into operation leading to a *gauche–gauche* preference about the two central C—O bonds (1).

O

g

O g

(1)

There is unquestionably an apparent attractive interaction between oxygen atoms situated 1,4 in a carbon chain which leads to a small preference of ca 0.4 kcal mol^{-1} for the *gauche* conformation in compounds such as 1,2-dimethoxyethane.[5] This preference has been shown to extend into polymeric systems.[6] As was pointed out in Chapter 2 there is nothing extraordinary about this *gauche* preference, which can be reproduced in related fluoro compounds by standard molecular mechanics calculations.[7] It must, however, be taken into account when dealing with compounds containing oxygen atoms in a 1,4-relationship.

Tetrahydropyran

Microwave studies show that tetrahydropyran (2) has a chair conformation.[8] This is confirmed by R value measurements which suggest that the ring is slightly flatter than cyclohexane.[9]

The free energy of activation for ring inversion is 10.3 kcal mol^{-1},[10,11] which is very similar to that in cyclohexane but less than in piperidine. It is interesting to note that the ring inversion barrier drops as one proceeds down the group VI pentamethylene heterocycles. The barriers are oxide, 10.3; sulphide, 9.4; selenide, 8.2; telluride, 7.3 kcal mol^{-1}.

(2)

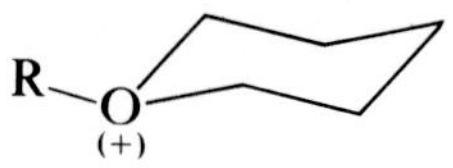

R—O
(+)

(3)

In principle, conformational equilibria on oxygen in oxonium salts (3) should be observable in a similar way to those in the piperidines. The results would of course be of great theoretical interest. In practice, however, the problem is much more difficult than in nitrogen-containing rings. Although the simple protonated form (3, R = H) has been prepared[11] the hydrogen on oxygen undergoes chemical exchange too rapidly to permit the determination of ^1H—^1H coupling constants and hence the conformational preference of the hydrogen on oxygen. The O-methyloxonium salt (3, R = Me) is similarly of interest for comparison with N-methylpiperidine. However, oxygen inversion in this type of system is extremely rapid and ring inversion could not be frozen out down to $-70°$ which was the limit of solubility.[12] Thus no information about conformational equilibria on oxygen was obtained.

Tetrahydropyran derivatives display the anomeric effect. Indeed as discussed in Chapter 2 it was from a study of pyranose sugars that the anomeric effect first became known. Briefly, the anomeric effect is the axial preference of an electronegative substituent, typically an alkoxy group, at position 2 in the ring (4). This preference varies from 0 to ca 2 kcal mol^{-1}, and generally displays a solvent dependence typical of dipolar interactions, with the more polar equatorial conformations becoming less unstable in solvents of higher dielectric constant.[13–16]

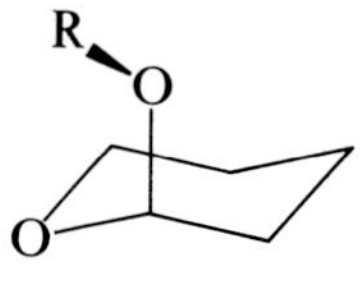

R—O

O

(4)

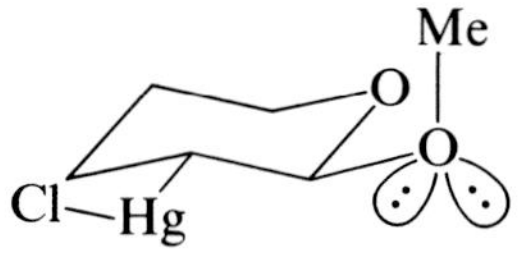

Me
O
|
O
Cl—Hg

(5a)

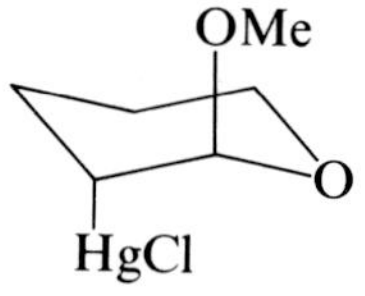

OMe

O

HgCl

(5b)

Although 2-alkoxy- and 3-mercuri-substituents prefer to be axial in a tetrahydropyran ring, these tendencies are overcome in the 2-methoxy-3-chloromercuri derivative (5) in which the diequatorial conformation (5a) is preferred.[17] This presumably arises from a more favourable oxygen–mercury interaction in (5a), where a pair of electrons on the methoxyl is directed towards the mercury, than in (5b) where this is not possible.

The anomeric effect persists in 1,4-dioxan derivatives (6),[18–22] but disappears next to oxygen in 1,4-oxathian derivatives (7).[23] It has been

(6)   (7)

suggested[23] that this arises from a negation of the anomeric effect by the "*gauche* repulsive effect",[24, 25] which occurs between the larger electronegative atoms when placed *gauche* about a single bond.

Conformational equilibria in 3-chloro- and 3-bromo-tetrahydropyran (8) have been studied.[26a] The equilibria are markedly solvent dependent, as would be expected if dipolar factors were important, with the equatorial conformation predominating. The free energy difference in the non-polar solvent carbon tetrachloride is 0.68 kcal mol^{-1} (8, X = Cl) at 20°, compared

(8eq)   (8ax)

with 0.53 kcal mol^{-1} in chlorocyclohexane (Chapter 1, Table 1.2). The equatorial preference is somewhat greater in the bromo compound (8, X = Br). The magnitudes of these energy differences are interesting. Firstly, the 3-axial substituent is *syn*-axial to only one hydrogen atom and consequently the compound should have a smaller preference for equatorial. For instance, a methyl group at the 3-position of the tetrahydropyran ring has a smaller preference for equatorial (1.27 kcal mol^{-1}) compared to 1.7 kcal mol^{-1} in cyclohexane or at the 2 and 4 positions in tetrahydropyran.[27] However, acting in the opposite direction should be a dipole–dipole interaction favouring the equatorial conformation. The combination of these effects must give rise to the observed free energy difference.

Anderson and Sepp have studied the conformational equilibria of methyl groups at positions 2, 3, and 4 in a tetrahydropyran ring by equilibration of *cis*- and *trans*-isomers of 4, 5, and 6-methyl-2-carbomethoxytetrahydropyran. This work also gave information on the conformational equilibrium of a 2-carbomethoxy group. Values for ΔG at 25° in methanol are 2-Me, 1.70 ± 0.10; 3-Me, 1.27 ± 0.05; 4-Me, 1.70 ± 0.10; 2-CO$_2$Me, 1.62 ± 0.05 kcal mol^{-1}.*

---

* More recent work, however, suggests these values to be in error. Eliel and Pietrusiewicz find ΔG^0 for *trans*-2,4-dimethyltetrahydropyran to be 0.89 kcal mol^{-1} at −100° favouring 2-Me equatorial (personal communication). See also E. Kleinpeter, C. Duschek and M. Muehlstaedt (1978). *J. Prakt. Chem.* **320**, 303.

The values for methyl fall in line with expectations outlined in Chapter 2. The value for the carbomethoxy group is larger than in cyclohexane (ca 1.2 kcal mol^{-1}) and may arise as an example of the "reverse anomeric effect".

1,2-Dioxan

Although 1,2-dioxans are a fairly well investigated series of compounds synthetically, little conformational work seems to have been done in this series. Their conformational analysis is expected to be dominated by the large two-fold potential energy function for rotation about the O—O bond (Chapter 2). This is borne out to some extent by photoelectron spectra of the 3,3,6,6-tetramethyl derivative (9) which suggest a large dihedral angle of 80–90° about the O—O bond.[28] A similar indication comes from the substantial barrier to ring inversion in 3,3,5-trimethyl-1,2-dioxene (10) of 10.5 kcal mol^{-1}, which is much larger than in cyclohexene.[29]

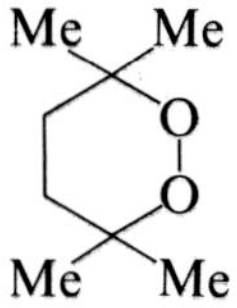

Me Me   Me Me

     O   O

     O   O

Me Me   Me

  (9) 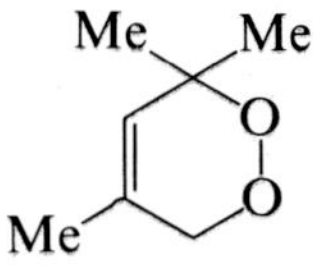   (10)

1,3-Dioxan

More is known about the conformational analysis of 1,3-dioxan than about that of any other heterocyclic system. This arises for three principal reasons. First, 1,3-dioxans are very readily synthesized by condensing any one of a wide variety of easily prepared 1,3-diols with aldehydes or ketones. Secondly, 1,3-dioxans show very readily interpretable NMR spectra because of the differential shielding introduced by the two ring oxygen atoms. Finally the presence of an acetal or ketal grouping in the ring allows very ready acid-catalysed epimerization equilibria to be established via open-chain intermediates. These equilibrations can serve as models for conformational equilibria. Consequently, interest in 1,3-dioxans has been considerable, and a large part of our knowledge of the conformational behaviour of reduced heterocyclic systems arises from studies of this ring system. In addition, many special conformational effects for heterocyclic systems are well illustrated by the 1,3-dioxans.

Several crystallographic studies[30-32] and an electron diffraction investigation[33] have been published. All are of molecules in chair conformations, including even the extremely hindered molecule (11),[30] indicating the very strong preference of the 1,3-dioxan ring for the chair conformation.

The ring geometries of the compounds (12), (13)[31] and (14)[32] are very similar. The unstrained rings (12) and (14) are appreciably more puckered about the oxygen atoms (O(1), C(2), O(3)) where the dihedral angles are 60–63° than about the alicyclic portion of the ring (C(4), C(5), C(6)) where the dihedral angles are 53–55°—similar to those in cyclohexane. It is of interest that the C–C bond lengths (149–151 pm) are significantly shorter than in cyclohexane (153.5 pm), making for a more compact ring.

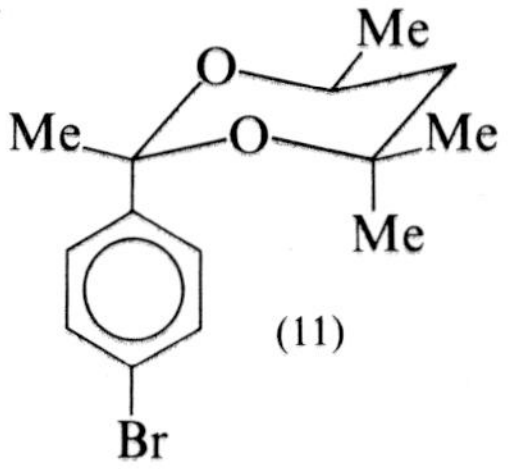

Me $\quad$ (11)

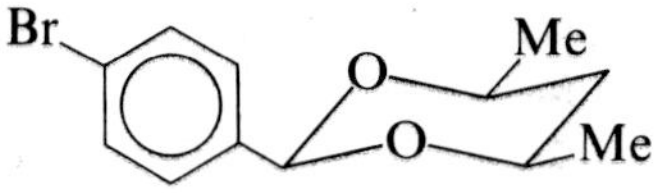

Br $\quad$ (12)

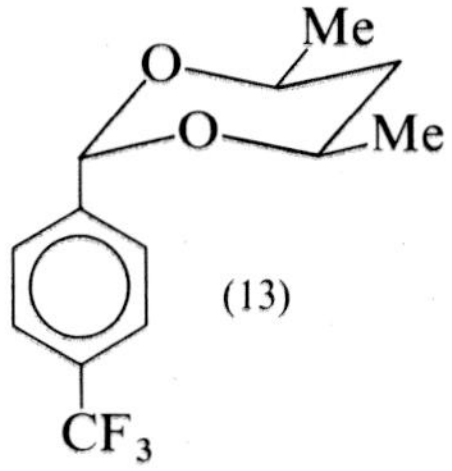

Me $\quad$ (13)

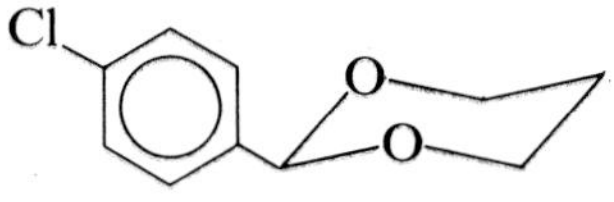

Cl $\quad$ (14)

For the axial phenyl derivative (13) the strain arising from the axial 2-phenyl group is relieved in several ways. The phenyl leans away from the undistorted axial position by ca 7°. The puckering in O(1), C(2), O(3) region of the molecule is considerably reduced compared to (12) or (14) making all the ring dihedral angles 53–55° and the ring somewhat flatter. Finally the ring atom C(2) is displaced from its expected position roughly in the plane of the benzene ring.

The severely hindered compound (11) was synthesized and examined in the hope of observing a non-chair conformation of a 1,3-dioxan ring.[30] Despite the severe strain present in (11), as indicated by the 2-p-bromophenyl leaning some 30° out of a true axial position the molecule still adopts a chair conformation.

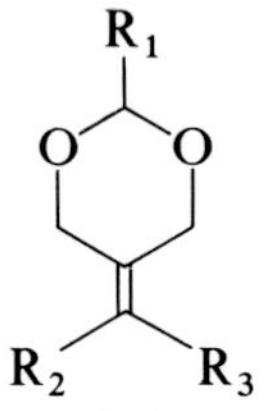

$$R_1$$

$$R_2 \qquad R_3$$

(15)

Although 1,3-dioxan seems from the above studies to have a more puckered ring than cyclohexane, NMR results from Anteunis seem to show that introduction of a 5-alkylidene group (15) flattens the ring to a greater extent than in cyclohexane.[34]

Chemical equilibrations have proved of inestimable value in the conformational study of 1,3-dioxans. They can be of two types: epimerization or exchange reactions. In the epimerization reactions a Lewis acid or an anhydrous protic acid catalyses a ring opening–closing reaction (Fig. 4.1). Although this reaction formally involves epimerization at C(2), subsequent ring inversion of the new diastereoisomer effectively causes epimerization at other ring positions. In the exchange reactions the acid catalyses the exchange of ketone (or aldehyde) and diol parts of two different compounds, thus setting up a four-component equilibrium (Fig. 4.2). This exchange technique, although more difficult to carry out than the simpler equilibrations, can yield information about some otherwise inaccessible conformational equilibria.

The first equilibrations on 1,3-dioxans were reported by Riddell and Robinson[35] and by Eliel and Knoeber.[36] It soon became apparent that a methyl group axial in the 2-position is considerably more hindered than in cyclohexane (ΔG° 2-Me ca 4.0 kcal mol^{-1}). An axial group in the 4-position is somewhat less hindered, but again more hindered than in cyclohexane (ΔG° 4-Me ca 3.1 kcal mol^{-1}), and a methyl group in the 5-position is considerably less hindered when axial than in cyclohexane (ΔG° 5-Me ca 0.9 kcal mol^{-1}). From these and other studies that have measured enthalpy and entropy differences, it appears that entropy plays a minor role in deter-

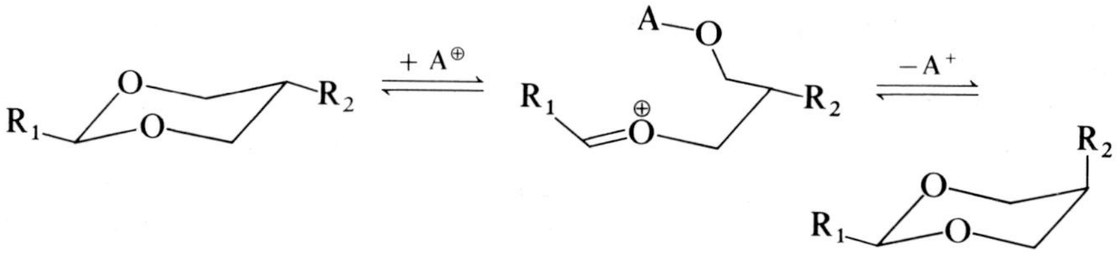

Fig. 4.1   Acid-catalysed epimerization of a 1,3-dioxan derivative.

mining the free energy difference, which is almost entirely dominated by the enthalpy term. It is also apparent that solvent changes can cause variations in the free energy difference. This solvent dependence is understandable in this system because differential solvation of the oxygen and methylene groups will vary with the solvent employed.

The variation in $\Delta G°$ for methyl at the three ring positions is readily understood when the known structure of the ring discussed earlier is considered. Because of the increased puckering at the C(2) end of the molecule and the short C—O bond lengths the axial group at C(2) is brought much closer to the axial 4,6 hydrogen atoms than might otherwise have been expected (16). This increases the repulsive interactions raising the free energy difference. The interactions on an axial 4-substituent are less severe because one of the *syn*-axial hydrogens, on C(6), is now more distant and the axial group is no longer forced into the ring by the puckering (17). An axial group at position 5 has non-bonded interactions only with oxygen atoms (18).

(16)   (17)   (18)

Since oxygen is appreciably smaller than methylene (Table 2.1) this interaction is reduced considerably compared to positions 2 and 4 or compared to cyclohexane. Moreover, the axial 5-position appears to lean out of the ring to an appreciable extent reducing the interactions even further.

Interestingly it is found possible to place a *t*-butyl group axial at position 5 in the ring ($\Delta G° = 1.4–1.9$ kcal mol^{-1}) without forcing the ring into a twist conformation as would be expected in cyclohexane and several other six-membered heterocyclic systems.[35] In fact the chair–twist free energy difference in 1,3-dioxan is the highest that has yet been determined experimentally (see below).

There have been extensive subsequent reports of equilibrations to investigate conformational equilibria in 1,3-dioxans notably by the Belgian group lead by Anteunis and by the American group lead by Eliel.[37] It is beyond the scope of this chapter to cover all results in this field and consequently only those results of direct relevance to the theory of heterocyclic conformational analysis will be discussed in detail (see Table 4.1).

The distinct tendency of the small electronegative atoms oxygen and fluorine when arranged 1,4 in a chain, to get into a *gauche* arrangement is

TABLE 4.1 Conformational free energy, enthalpy and entropy differences for substituents on a 1,3-dioxan ring from equilibration studies.

Group	ΔG^b	T	ΔH (kcal mol^{-1})	ΔS (cal-K^{-1} mol^{-1})	Solvent[a]	Ref.
2-Me	3.98	25	—	—	Et$_2$O	50
2-Et	4.04	25	—	—	Et$_2$O	50
2-i-Pr	4.17	25	—	—	Et$_2$O	50
2-Ph	3.12	25	—	—	Et$_2$O	50
2-OMe	−0.36 to −0.50	25	—	—	Et$_2$O	50
4-Me	2.9	25	—	—	Et$_2$O	36
5-Me	0.80	25	—	—	Et$_2$O	36
	0.89	25	0.86	−0.1	CHCl$_3$	35
5-Et	0.67	25	—	—	Et$_2$O	36
	0.81	25	0.74	−0.2	CHCl$_3$	35
5-i-Pr	0.98	25	—	—	Et$_2$O	36
	1.10	25	1.13	+0.1	CHCl$_3$	35
5-t-Bu	1.36	25	1.54	+0.6	Et$_2$O	36
	1.7	25	1.9	+0.5	CHCl$_3$	35
5-Ph	1.03	25	—	—	Et$_2$O	36
5-OH	−0.915	80	—	—	C$_6$H$_{12}$	38
	−0.51	80	—	—	iPrOH	38
	−0.50	80	—	—	tBuOH	38
	−0.27	80	—	—	di(MeO)Et	38
5-OMe	0.89	28	—	—	CCl$_4$	38
	0.83	50	—	—	Et$_2$O	38
	0.18	28	—	—	CHCl$_3$	38
	0.03	28	—	—	CH$_3$OH	38
	−0.01	25	—	—	CH$_3$CN	38
5-OAc	0.0	25	—	—	Et$_2$O	38
5-F	−0.62	25	—	—	Et$_2$O	38
	−0.605	25	—	—	CH$_3$OH	38
	−0.83	25	—	—	C$_6$H$_6$	38
	−1.225	25	—	—	CH$_3$CN	38
5-Cl	0.94	25	—	—	CHCl$_3$	40
	1.20	25	—	—	Et$_2$O	47
5-Br	1.35	25	—	—	CHCl$_3$	40
	1.44	25	—	—	Et$_2$O	47
5-I	1.43	20	—	—	CHCl$_3$	48
5-NO$_2$	−0.38	30	—	—	CCl$_4$	38
	−0.63	30	—	—	CHCl$_3$	38
	−0.81	30	—	—	CH$_2$Cl$_2$	38
	−0.89	30	—	—	CDCl$_3$	38
5-CN	0.21	30	—	—	Et$_2$O	38
	−0.55	30	—	—	CH$_3$CN	38
5-CO$_2$Me	0.82	30	—	—	Et$_2$O	38
	0.22	27	—	—	CH$_3$CN	38

TABLE 4.1 (Continued)

Group	ΔG^b	T	ΔH (kcal mol^{-1})	ΔS (cal-K^{-1} mol^{-1})	Solvent^a	Ref.
5-CH$_2$OH	−0.27	27	—	—	CCl$_4$	38
	+0.01	30	—	—	di(MeO)Et	38
5-CH$_2$OCH$_3$	0.05	30	—	—	Et$_2$O	38
5-SMe	1.74	26.5	—	—	CCl$_4$	47
	1.55	26.5	—	—	C$_6$H$_6$	47
5-SOMe	−0.74	54	—	—	C$_6$H$_6$	47
	−0.82	54	—	—	CHCl$_3$	47
5-SO$_2$Me	−1.07	50	—	—	C$_6$H$_6$	47
	−1.19	50	—	—	CHCl$_3$	47
5-SMe$_2^+$	−2.0	25	—	—	TFA	47

^{a} Solvents: Et$_2$O = diethylether, C$_6$H$_{12}$ = cyclohexane, di(MeO)Et = 1,2-dimethoxy-ethane, TFA = trifluoracetic acid.
^{b} Positive energies indicate that equatorial is more stable.

illustrated by the axial preference of a 5-fluoro group (19).[38–40] This prefer-ence is solvent dependent, showing the greatest axial tendency in polar solvents. The theory of variation of conformer energy with solvent dielectric constant developed by Abraham and Gatti,[41] gives good agreement with the experimental results.[40] The axial preference of a 5-alkoxy group is consider-ably smaller than that of a 5-fluoro group, but shows a similar solvent dependence.[38,42,43] Indeed it is only in the most polar solvents that the axial conformation is marginally preferred (20).

A 5-hydroxyl group has an axial preference in all solvents studied, pre-sumably because of hydrogen bonding between the hydroxyl group and the ring oxygen atoms (21).[38] In accordance with this idea, the preference is greater in cyclohexane than in other solvents capable of accepting a hydrogen bond from the hydroxyl group. It has been suggested that the axial hydroxyl group in *cis*-2-phenyl-5-hydroxy-1,3-dioxan (21) has a bifurcated hydrogen bond from the hydroxyl group to both ring oxygen atoms.[44] Similarly a 5-amino substituent has been shown to have an axial preference, again presumably for hydrogen bonding reasons.[45]

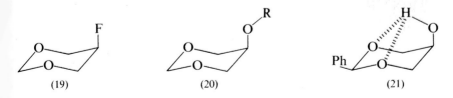

(19)   (20)   (21)

The trimethyl ammonium group at position 5 prefers to be axial (22).[46] This is another illustration of charge–dipole interactions dominating conformational behaviour in heterocyclic systems. Further illustrations are found in the comparison of 5-SMe, 5-SOMe, and 5-SO_2Me groups.[47] In the case of the thiomethyl group (23) the *equatorial* preference is ca 1.7 kcal mol^{-1} in a variety of solvents. For the sulphone (24) the *axial* preference is ca 0.8 kcal mol^{-1} in the same solvents, whilst for the sulphoxide (25) the *axial* preference is ca 1.1 kcal mol^{-1}. The methyl sulphonium salt (26) shows a greater axial preference. These results probably arise from an attraction between the electronegative oxygen atoms and the formal (or real) positive charge on the sulphur.

The free energy difference of the 5-thioether (23) (ca 1.7 kcal mol^{-1}) is greater than in cyclohexane (1.07 kcal mol^{-1}) and provides another example of the *gauche* repulsive effect.

(22)	(23)	(24)

(25)	(26)	(27)

The *gauche* repulsive effect is probably also operative in the 5-halogeno-1,3-dioxans. Taking values for chloroform solutions the equatorial preferences increase as the size of the halogen increases[47,48] Cl, 0.94; Br, 1.35; I, 1.43 kcal mol^{-1}. This increase contrasts with the values for the halocyclohexanes (Table 1.2) where the values are all similar at ca 0.4–0.5 kcal mol^{-1}.

The 5-hydroxymethyl group has a small axial preference (27) (+0.01 to −0.27 kcal mol^{-1})[47] but surprisingly shows no evidence of hydrogen bonding between the ring oxygen atoms and the hydroxyl group. This interesting observation was investigated by Eliel and Banks,[49] who concluded that an interplay between dipolar, bond eclipsing and hydrogen bonding forces had to be invoked to account for observations in this series.

Despite the extremely severe steric hindrance of axial substituents at position 2 in a 1,3-dioxan ring, it is found that a 2-methoxyl group has a small (ca 0.4 kcal mol^{-1}) preference to be axial.[50] This is a reflection of the reinforced anomeric effect provided at the 2-position by both ring oxygen atoms overcoming the severe steric constraints.

The chair–twist free energy difference in 1,3-dioxan is ca 8 kcal mol^{-1} and consequently is much too large to be amenable to study by the conventional techniques of conformational analysis such as two component equilibration because of the minute amount of the minor component that would be present at equilibrium. As a consequence, measurement of this parameter posed a considerable problem and challenge for several years and prompted the introduction of new techniques specifically to deal with this problem. These techniques may subsequently prove useful in the solution of other problems.

Firstly molecular rotation[51] and NMR measurements[50, 52, 53] established that certain twist conformations, e.g. (28), were preferred over the corresponding chair conformations, e.g. (29), which would have 2,4-*syn*-diaxial methyl groups. Having established this, two techniques, microcalorimetry and four-component equilibration, were employed to obtain the energy difference. Microcalorimetric measurement of the enthalpy change for the acid-catalysed reaction (28) → (30) gives the twist → chair change in this system.[54]

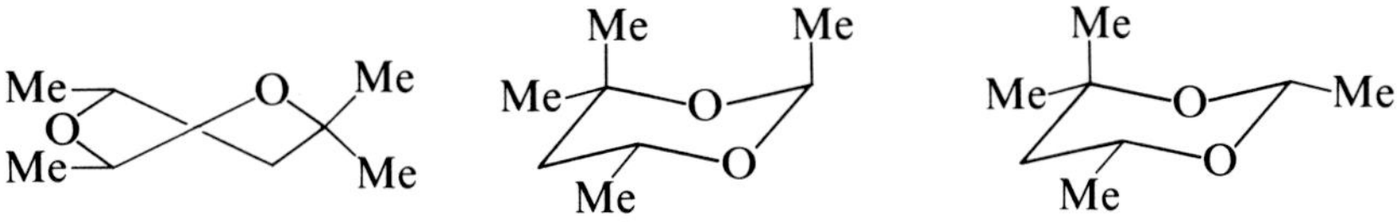

(28)   (29)   (30)

The appropriate corrections for strain in (28) and (30) and for entropy gave a value for the free energy difference between the twist and chair forms of ca 8.0 kcal mol^{-1}. Establishment of the four component equilibrium displayed in Fig. 4.2 also allows an estimation of the chair–twist free energy difference.[55] Compound (31) is known to have a twist conformation whilst the others have chairs. Summation of the strain known to be present in (33)

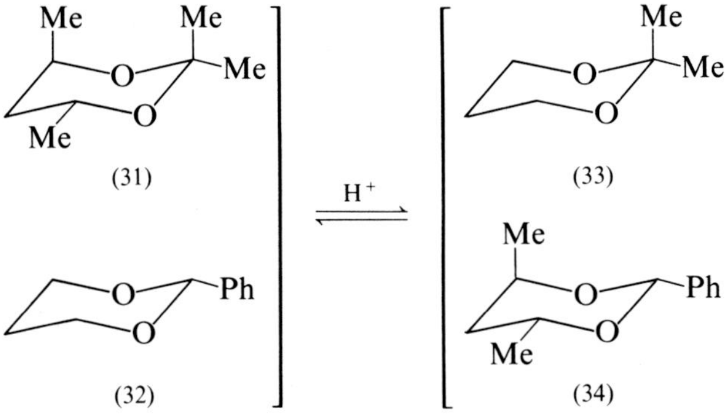

(31)

(32)

H$^+$

(33)

(34)

Fig. 4.2 Acid-catalysed four-component exchange reaction in 1,3-dioxan derivatives.

and (34) due to the presence of axial alkyl groups, with the experimental free energy difference gives a chair–twist free energy difference of 7.4 kcal mol^{-1}.

The problem has also been tackled in the vapour phase by appearance potential measurements in the mass spectrum[56] and by ultrasonic relaxation measurements in the liquid phase.[57] Other probably less reliable measurements have been reported.[58,59]

From these studies of the twist form of 1,3-dioxan several interesting and general points worth noting arise. 1,3-Dioxan may exist in two twist conformations which are known as the 1,4- (35) or 2,5- (36) twists depending upon which atoms in the ring lie along the (*quasi*) two-fold axis of the molecule. The microcalorimetric work described above therefore refers to 1,4 twist → chair change, whereas the four-component equilibration refers to the 2,5 twist → chair change. The extremely large values obtained for the free energy changes probably arise from the compact nature of the ring which will exacerbate non-bonded interactions in twist conformations.

(35)   (36)

Prior to this work it was believed that twist conformations could pseudo-rotate or pseudolibrate fairly freely. The observation that certain 1,3-dioxans occupy fairly rigid and well-defined twist conformations casts doubt on these ideas. Indeed if a molecule is so encumbered as to be forced into a twist conformation it is very probable that the encumbrance will persist in the twist form, severely restricting the number of twist conformations available.[60]

There have been several studies of the barrier to ring inversion in 1,3-dioxan.[62] The barrier in the parent compound is found to be ca 9.7 kcal mol^{-1}.[61,62] Theoretical calculations of ring inversion barriers and pathways in a variety of oxygen-containing rings have been performed by Pickett and Strauss.[63]

1,4-Dioxan

1,4-Dioxan has been much less intensively studied than 1,3-dioxan. Little is known about conformational equilibria in this system although fairly extensive structural data from diffraction methods exist in the literature.[64-72]

In 1,4-dioxan itself, electron diffraction shows the molecule to have a chair conformation that is slightly more puckered than cyclohexane with average internal ring dihedral angles of 57.9°. The introduction of halogens onto the ring carbon atoms causes several interesting distortions of the ring shape. First the anomeric effect operates and the halogens, wherever possible, go into axial positions. Secondly, in *trans*-2,3-dihalo-1,4-dioxans such as *trans*-2,3-dichloro[66] or dibromo [65]-1,4-dioxan (37) the ring dihedral angle at the halogen side decreases flattening the ring and the dihedral angle between the two axial halogens decreases from 180°. This effect is most pronounced in *trans-syn-trans*-tetrachloro-1,4-dioxan (38).[69] Although the halogens wish to become axial due to the anomeric effect the axial halogen at position 2 has a *gauche* relationship with the oxygen at position 4 in the ring introducing the *gauche* repulsive effect. The observed distortions minimize the *gauche* repulsions whilst maintaining the favourable axial anomeric interaction.

X = Cl or Br

(37)   (38)

The measurement of the ring inversion barrier in 1,4-dioxan proved a problem for a considerable time because of the extremely small chemical shift difference between the axial and equatorial hydrogen atoms in the slow exchange limit spectrum. This problem was eventually solved by two groups.[73–75] Jensen's group made use of changes in ^{13}C satellite spectra in the 100 MHz ^{1}H spectrum.[73,74] Anet's group[75] made use of hexadeuterio-1,4-dioxan and measured a chemical shift difference of 1.6 Hz (at 100 MHz) at the slow exchange limit. Both groups measured the free energy of activation to be 9.7 kcal mol^{-1}.

Several groups have reported work on the 1,4,5,8-tetraoxadecalin system (39). One of the most interesting features is the complete absence of any reports of the *trans*-fused system (40). In the *cis*-fused ring there are three attractive *gauche* 1,4-dioxygen interactions about the central bond and two "anomeric" axial oxygen atoms. In the *trans*-isomer there are only two *gauche* relationships of oxygen atom pairs. Furthermore much of the instability associated with the *cis*-decalin structure is removed by the *endo* oxygen atoms, with their reduced non-bonded interactions with the methylene groups in the opposite ring. Fuchs and his group have reported on some conformational equilibria in this ring system and on the ring inversion barrier ($\Delta G^{\ddagger} = 11.2$ kcal mol^{-1}).[70,76]

The ring inversion barrier in the tricyclic derivative (41) has been found to be somewhat higher than in the bicyclic derivative at 12.6 kcal mol^{-1}.[77]

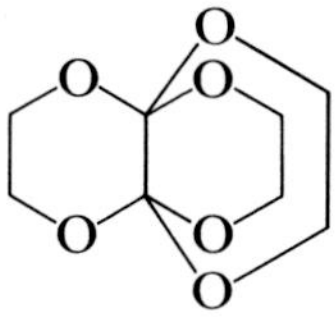

(41)

1,3,5-Trioxan

Although structural data are available from microwave,[78] X-ray,[79,80] and electron diffraction[81,82] studies, and the 1,3,5-trioxan ring is found to have a well-defined chair conformation, little work has been reported on the conformational equilibria in this system.

1,2,4,5-Tetraoxan

X-Ray diffraction studies of several derivatives of this ring have been reported.[83] The anticipated enlargement of the ring dihedral angle about the O—O bond is observed.[84]

References

1. E. L. Eliel. (1970). *Acc. Chem. Res.* **3**, 1.
2. S. Wolfe, A. Rauk, L. M. Tel and I. G. Csizmadia. (1971). *J. Chem. Soc. (B)* 136.
3. J. Dale. (1974). *Tetrahedron* **30**, 1683.
4. N. L. Allinger and D. Y. Cheung. (1976). *J. Am. Chem. Soc.* **98**, 6798.
5. R. Iwamoto. (1971). *Spectrochim. Acta* **27A**, 2385.
6. K. Matsuzaki and H. Ito. (1974). *J. Polym. Sci. Polym. Phys. Ed.* **12**, 2507.
7. R. J. Abraham and P. Loftus. (1974). *J.C.S. Chem. Comm.* 180.
8. V. M. Rao and R. Kewley. (1969). *Canad. J. Chem.* **47**, 1289.
9. J. B. Lambert. (1971). *Acc. Chem. Res.* **4**, 87.

10. J. B. Lambert, C. E. Mixan and D. H. Johnson. (1973). *J. Am. Chem. Soc.* **95**, 4634.
11. J. B. Lambert, R. G. Keske and D. K. Weary. (1967). *J. Am. Chem. Soc.* **89**, 5921.
12. J. B. Lambert and D. H. Johnson. (1968). *J. Am. Chem. Soc.* **90**, 1349.
13. C. B. Anderson and D. T. Sepp. (1968). *Tetrahedron* **24**, 1707.
14. A. J. de Hoog, H. R. Buys, C. Altona and E. Havinga. (1969). *Tetrahedron* **25**, 3365.
15. G. O. Pearson and O. A. Runquist. (1968). *J. Org. Chem.* **33**, 2572.
16. E. L. Eliel and C. A. Giza. (1968). *J. Org. Chem.* **33**, 3754.
17. N. S. Zefirov and N. M. Schechtman. (1967). *Dokl. Acad. Sci. SSSR* **177**, 842.
18. C. Altona, C. Romers and E. Havinga. (1959). *Tetrahedron* **10**, 16.
19. C. Altona, C. Knobler and C. Romers. (1963). *Acta Cryst.* **16**, 1217.
20. C. Altona, C. Knobler and C. Romers. (1963). *Rec. Trav. Chim.* **82**, 1089.
21. C. Altona and C. Romers. (1963). *Rec. Trav. Chim.* **82**, 1080.
22. N. S. Zefirov and M. A. Fedorovskaya. (1969). *Zh. Org. Khim.* **5**, 158.
23. N. S. Zefirov, V. S. Blagoveshchensky, I. V. Kozimirchik and N. S. Surova. (1971). *Tetrahedron* **27**, 3111.
24. N. S. Zefirov. (1977). *Tetrahedron* **33**, 3193.
25. E. L. Eliel and E. Juaristi. (1978). *J. Am. Chem. Soc.* **100**, 6114.
26a. C. B. Anderson and M. P. Geis. (1975). *Tetrahedron* **31**, 1149.
26b. R. Schrooten, F. Borremans and M. Anteunis. (1978). *Spectrochim. Acta A* **34A**, 297.
27. C. B. Anderson and D. T. Sepp. (1968). *J. Org. Chem.* **33**, 3272.
28. C. Batich and W. Adam. (1974). *Tetrahedron Lett.* 1467.
29. M. L. Kaplan and G. N. Taylor. (1973). *Tetrahedron Lett.* 295.
30. G. M. Kellie, P. Murray-Rust and F. G. Riddell. (1972). *J.C.S. Perkin II* 2384.
31. F. W. Nader. (1975). *Tetrahedron Lett.* 1207, 1591.
32. A. J. de Kok and C. Romers. (1970). *Rec. Trav. Chim.* **89**, 313.
33. G. Schultz and I. Hargittai. (1974). *Acta Chim. Acad. Sci. Hung* **83**, 331.
34. M. Anteunis and R. Camerlynk. (1975). *Tetrahedron* **31**, 1841.
35. F. G. Riddell and M. J. T. Robinson. (1967). *Tetrahedron* **23**, 3419.
36. E. L. Eliel and M. C. Knoeber. (1968). *J. Am. Chem. Soc.* **90**, 3444.
37. For some leading references see Ref. 1 in this chapter, and also E. L. Eliel. (1972). *Angew. Chem.* (Int. edn) **11**, 739.
38. E. L. Eliel and M. K. Kaloustian. (1970). *Chem. Comm.* 290.
39. L. D. Hall and R. N. Johnson. (1972). *Org. Mag. Res.* **4**, 369.
40. R. J. Abraham, H. D. Banks, E. L. Eliel, O. Hofer and M. K. Kaloustian. (1972). *J. Am. Chem. Soc.* **94**, 1913.
41. R. J. Abraham and G. Gatti. (1969). *J. Chem. Soc.* (*B*) 961.
42. B. J. Hutchinson, R. A. Y. Jones, A. R. Katritzky, K. A. F. Record and P. J. Brignall. (1970). *J. Chem. Soc.* (*B*) 1224.
43. E. L. Eliel and O. Hofer. (1973). *J. Am. Chem. Soc.* **95**, 804.
44. J. C. Jochims and Y. Kobayashi. (1976). *Tetrahedron Lett.* 2065.
45. E. Coene and M. Anteunis. (1970). *Tetrahedron Lett.* 595.
46. R. Van Cauwenberghe, M. Anteunis and L. Valcky. (1974). *Bull. Soc. Chim. Belges* **83**, 285.
47a. M. K. Kaloustian, N. Dennis, S. Mager, S. A. Evans, F. Alcudia and E. L. Eliel. (1976). *J. Am. Chem. Soc.* **98**, 956.
47b. E. L. Eliel, D. Kandasamy and R. C. Sechrest. (1977). *J. Org. Chem.* **42**, 1533.
48. N. J. Kotite, M. Harris and M. K. Kaloustian. (1977). *J.C.S. Chem. Comm.* 911.
49. E. L. Eliel and H. D. Banks. (1972). *J. Am. Chem. Soc.* **94**, 171.
50. F. W. Nader and E. L. Eliel. (1970). *J. Am. Chem. Soc.* **92**, 3050.

51. J. F. Tocanne. (1970). *Bull. Soc. Chim. France* 750.
52. G. M. Kellie and F. G. Riddell. (1971). *J. Chem. Soc.* (*B*) 1030.
53. K. Pihlaja, F. G. Riddell and G. M. Kellie. (1972). *J.C.S. Perkin II* 252.
54. R. M. Clay, G. M. Kellie and F. G. Riddell. (1973). *J. Am. Chem. Soc.* **95**, 4632.
55. E. L. Eliel, J. R. Powers and F. W. Nader. (1974). *Tetrahedron* **30**, 515.
56. K. Pihlaja and J. Jalonen. (1971). *Org. Mass Spectrom.* **5**, 1363.
57. G. Eccleston and E. Wynn-Jones. (1971). *J. Chem. Soc.* (*B*) 2469.
58. K. Pihlaja. (1968). *Acta Chem. Scand.* **22**, 716.
59. M. Anteunis and G. Swaelens. (1970). *Org. Mag. Res.* **2**, 389.
60. For further discussion of these points see: G. M. Kellie and F. G. Riddell (1974). *In* "Topics in Stereochemistry", Vol. 8, (Ed. Eliel and Allinger), 225. Wiley Interscience, New York.
61. For leading references see: J. E. Anderson. (1974). *Fortschritte* **45**, 139.
62. G. Binsch, E. L. Eliel and S. Mager. (1973). *J. Org. Chem.* 4079.
63. H. M. Pickett and H. L. Strauss. (1970). *J. Am. Chem. Soc.* **92**, 7281.
64. For an excellent review of the literature up to 1969 see: C. Romers, C. Altona, H. R. Buys and E. Havinga. (1969). *In* "Topics in Stereochemistry", Vol. 4, (Eds Eliel and Allinger), 39. Wiley Interscience, New York.
65. C. Altona, C. Knobler and C. Romers. (1963). *Rec. Trav. Chim.* **82**, 1089.
66. C. Altona and C. Romers. (1963). *Rec. Trav. Chim.* **82**, 1080.
67. C. Altona, C. Knobler and C. Romers. (1963). *Acta Cryst.* **16**, 1217.
68. C. Altona and C. Romers. (1963). *Acta Cryst.* **16**, 1225.
69. E. W. M. Rutten, N. Nibbering, C. H. MacGillavry and C. Romers. (1968). *Rec. Trav. Chim.* **87**, 888.
70. B. Fuchs, I. Goldberg and U. Shmueli. (1972). *J.C.S. Perkin II* 357.
71. M. Senona, Z. Tiari, K. Osaki and T. Taga. (1973). *J.C.S. Chem. Comm.* 880.
72. M. Davis and O. Hassel. (1963). *Acta Chem. Scand.* **17**, 1181.
73. F. R. Jensen and R. A. Neese. (1971). *J. Am. Chem. Soc.* **93**, 6329.
74. F. R. Jensen and R. A. Neese. (1975). *J. Am. Chem. Soc.* **97**, 4345.
75. F. A. L. Anet and J. Sandstrom. (1971). *Chem. Comm.* 1558.
76. B. Fuchs, Y. Auerbach and M. Sprecher. (1974). *Tetrahedron* **30**, 437.
77. B. Fuchs, Y. Auerbach and M. Sprecher. (1972). *Tetrahedron Lett.* 2267.
78. T. Oka, K. Tsuchiya, S. Iwata and Y. Morino. (1964). *Bull. Chem. Soc. Jpn* **37**, 4.
79. V. Busetti, A. Del Pra and M. Mammi. (1969). *Acta Cryst.* **B25**, 1191.
80. V. Busetti, M. Mammi and G. Z. Carazzolo. (1963). *Z. Krist.* **119**, 310.
81. A. H. Clark and T. G. Hewitt. (1971). *J. Molec. Struct.* **9**, 33.
82. A. E. Astrup. (1973). *Acta Chem. Scand.* **27**, 1345.
83. P. R. Groth. (1967). *Acta Chem. Scand.* **21**, 2608, 2631, 2695.
84. F. G. Riddell and D. R. Slim, unpublished results calculated from data in Ref. 83.

5
Six-Membered
Nitrogen-Containing Rings

General Considerations

Six-membered saturated heterocyclic rings containing nitrogen, the simplest member of which is piperidine (1), occur extensively in natural products, most noticeably in the alkaloids.

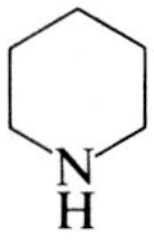

N
H

(1)

Conformational interest in these rings arises from two main factors. The nitrogen atom, in the free base, can invert changing the orientation of its substituent from equatorial to axial and vice versa. Also, the presence of the nitrogen atom in the ring will modify the positions of the conformational equilibria compared to cyclohexane. A carbon–nitrogen bond is somewhat shorter than a C–C bond (148 pm vs 153 pm) making nitrogen-containing rings more compact, but not as compact as oxygen-containing rings. The van der Waals radius of nitrogen is considerably smaller than that of a methylene group (ca 150 vs ca 200 pm) reducing non-bonded interactions involving the nitrogen atom.

The question of the barrier to nitrogen inversion in six-membered rings has been extensively discussed and disputed in the literature.[1-4] It is now agreed that the barrier to inversion of an equatorial N-alkyl group is raised by an adjacent electronegative atom (N or O), lowered by a β electronegative atom, and hardly affected by an electronegative atom in a γ position or further away.[3,4] These effects are understandable if one considers the way that the extra heteroatom affects the electron distribution on nitrogen. Quantum mechanical calculations show that an α electronegative atom should reduce the electron density on the reference atom, whilst a β electronegative atom increases it.[5] This is known as the "charge alternation" effect.

For an sp^3 hybridized nitrogen atom to invert it must pass through an sp^2 hybridized transition state with more s and less p character in the bonding orbitals. Any factor that increases the p character of the ground state will increase the barrier. Conversely any factor that decreases the p character of the ground state decreases the barrier. Since p orbitals have their electron densities further from the nucleus, electron-withdrawing substituents, such as α O or N, will increase the p character on nitrogen and increase the barrier. Similarly, electron-donating substituents, such as β O or N, will decrease the p character on nitrogen and reduce the barrier.

Barriers measured in the reverse direction, axial $\rightarrow$ equatorial, will differ from those in the forward direction only by the axial–equatorial free energy difference.[4]

As discussed in Chapter 2 the presence of α heteroatoms will introduce additional torsional interactions into the ring in compounds incorporating hydroxylamine or hydrazine moieties into the ring. The presence of heteroatoms β to nitrogen however will lower the energy of the axial N-alkyl conformation by removal of a *syn*-axial hydrogen and replacing it by a smaller group. It also turns out that "electrostatic" or anomeric interactions are lower in these compounds when the N-alkyl group is axial.

Piperidine

Crystallographic studies of salts of derivatives of piperidine[5-8] reveal that the ring adopts a chair conformation with torsion angles in the range 53–56°, very similar to those found for cyclohexane and its derivatives.[9] They also reveal a preference for substituents on nitrogen to be equatorial. For the derivative (2) of 3-hydroxypiperidine, the isomer observed, which is presumably the thermodynamically more stable, is *trans*, and the ester group at position 3 is axial.[6] Whilst the axial position of the 3-substituent

Et—N⊕
   |
   H   O—C—CH(Ph)$_2$
                ‖
                O
             (2)

may arise from crystal packing forces, it does allow a favourable electro-static interaction between the electronegative oxygen and the positive charge on the quaternary nitrogen atom as in acetylcholine. This may well therefore be the predominant conformation in solution as well (Chapter 2, p. 29).

There was for a considerable period an extended controversy in the litera-ture concerning the conformational equilibrium on nitrogen in piperidine.[10] In 1958 Aroney and Le Fevre, from a study of Kerr constants and molecular polarizability of certain piperidines in benzene concluded that "... the volume requirement of a lone electron pair on nitrogen exceeds that of a covalently bound hydrogen, and seems to approach in order of magnitude that of a methyl group . . .", an idea that had been foreshadowed by Barton and Cookson in their authoritative and influential review.[12] Le Fevre concluded that in the preferred conformation of piperidine the N—H was axial. The unfortunate wording of their statement, using the term "volume require-ment" and the novel implication it contained has subsequently led to much interesting and original work. Allinger[13] and Katritzky[14] studied the dipole moments of 4-p-chlorophenylpiperidine (3a) and its N-methyl derivative (3b), finding the equatorial conformations to be more stable by 0.4–0.5 kcal mol^{-1} (3a) and 1.7 kcal mol^{-1} (3b). Later dipole moment work[15] revised the latter value to 0.65 kcal mol^{-1}, although this value is now believed to be considerably too low (see below). Their conclusions were therefore contrary to those of Aroney and Le Fevre, who subsequently backed up their views by Kerr constant measurements on morpholine and N-methylmorpholine.[16]

Lambert and Keske compared the ^{1}H NMR spectra of piperidine and its N-methyl derivative.[17] The large chemical shift difference, between the C-(2, 6) hydrogens (ca 1 ppm) in the latter compound, known to have an equatorial N-methyl group, was attributed to an axial pair of electrons on

Cl—⟨benzene ring⟩—⟨piperidine ring⟩—NR

(3a) R = H
(3b) R = Me

the adjacent nitrogen atom. It was therefore argued that the smaller chemical shift difference in the former compound (ca 0.5 ppm) indicated an equatorial lone pair and hence an axial hydrogen atom. The same authors showed that protonated thian, selenan and telluran in FSO_3H/SO_2 at $-30°$ had axial hydrogen atoms.[18] The arguments here were more convincing as they arose from observations of one large and one small coupling constant to the adjacent methylene group. In a context such as this, chemical shifts provide much less reliable evidence than coupling constants, and consequently Lambert's arguments were criticized.[19, 20]

The problem has now finally been solved by three convincing pieces of work. The microwave spectrum of piperidine shows lines that can be attributed to the equatorial and axial conformations in which the equatorial predominates (ca 60% at 25°) and is thus of lower free energy in the vapour phase.[21]

An infrared spectral study of the first overtone region of the N—H stretching frequencies (ca $6500\ cm^{-1}$) of piperidine reveals two bands. From a consideration of the band outlines these were able to be assigned to the N—H equatorial and axial conformations.[22] The relative intensity of these bands varies with temperature. Since the extinction coefficients are unknown and cannot be presumed to be equal, the temperature variation leads only to values of the enthalpy difference between the conformations. The equatorial conformation is found to be lower in enthalpy by $0.5 \pm 0.1\ kcal\ mol^{-1}$.

Low-temperature NMR measurements on carefully dried piperidine (to avoid intermolecular exchange of the N—H) show two conformations at $-172°$ ($\Delta G° = 0.36\ kcal\ mol^{-1}$).[23] The first-order rate constant for nitrogen inversion (major → minor conformation) of $240 \pm 20\ s^{-1}$ at $-142°$ gives a free energy of activation $\Delta G^{\ddagger}$ of $6.1 \pm 0.2\ kcal\ mol^{-1}$.

There is thus now totally convincing evidence that both in the gas phase and in solution the N—H equatorial form is the more stable.

Although there never was much doubt that N-methyl and other N-alkyl piperidines preferred the N-equatorial conformations, there was for some time doubt as to the magnitude of the free energy difference. Earlier work[13, 14] based on dipole moment measurements suggested that the value for methyl was ca $1.7\ kcal\ mol^{-1}$, very similar to that in methylcyclohexane. This value was subsequently revised downwards by Katritzky et al.[15] The lower value was supported by the suggestion that an axial N-methyl group could lean out of the ring at a marginal cost in energy, because of the relatively low barrier to nitrogen inversion. This would alleviate non-bonded interactions with other ring groups and lower the free energy difference.

The latest and most convincing evidence on this topic comes from kinetically controlled trapping reactions at the nitrogen atom. Provided that a trapping reaction fulfills three conditions it can be used to investigate conformational equilibria. The reaction must: (a) be very much more rapid than the conformational interconversion; (b) proceed with a known stereochemical outcome, preferably retention of configuration; and (c) be irreversible on the timescale of the subsequent observations.[24] When these conditions are fulfilled the relative amounts of the products from the axial and equatorial conformations will correspond to the proportions of those conformations originally present. Robinson[25] used kinetically controlled protonation reactions to measure a value of 2.7 kcal mol^{-1}. McKenna[26] employed a kinetically controlled nitrene-trapping reaction to obtain a similar result. An analogous result is obtained for N,N'-dimethylpiperazine,[27] and the high value of ca 2.7 kcal is now generally accepted.[28]

The larger value for the conformational free energy difference in N-methylpiperidine compared to methylcyclohexane is best explained by the shorter C—N bonds bringing the axial N-methyl group closer to the syn-axial hydrogens than is possible in the carbocyclic case.

The conformational equilibrium in N-chloropiperidine has been investigated.[29] The free energy difference is found to be 1.5 ± 0.1 kcal mol^{-1}. Most interestingly, Anet's group have found the conformational free energy difference in N,N'-dichloropiperazine to be only 0.5 ± 0.1 kcal mol^{-1}.[30] Clearly there is either an anomaly here or some novel effect is manifesting itself.

Conformational equilibria of C-methyl groups at C(3) and C(4) in piperidine have been measured.[25b] As anticipated it is easier to put a methyl group axial at C(3) (1.51 ± 0.07 kcal mol^{-1}) than at C(4) (1.99 ± 0.07 kcal mol^{-1}). In the former case the value is lower than in cyclohexane because one syn-axial hydrogen atom has been removed. In the latter, the value is slightly higher than in cyclohexane possibly because of "pinching in" on carbon atoms 2 and 6. These values are largely substantiated by a more recent extensive survey by Eliel's group.[25c]

Hydrogen-1 NMR studies indicate that 2-alkyl groups in some 2-alkyl-N-nitropiperidine derivatives, exist in the axial orientation. This is because interference between the planar conjugated N-nitro group and the 2-substituent is much more severe when the 2-alkyl group is equatorial.[31]

Hexahydropyridazine

The hexahydropyridazine system is one of moderate complexity.[32] The ring contains a nitrogen–nitrogen bond which alters the torsional forces in this part of the ring and raises the nitrogen inversion barriers to a point where they are comparable with that for ring inversion.

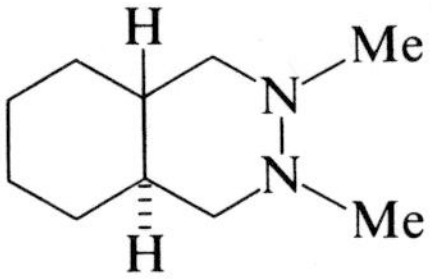

(4) 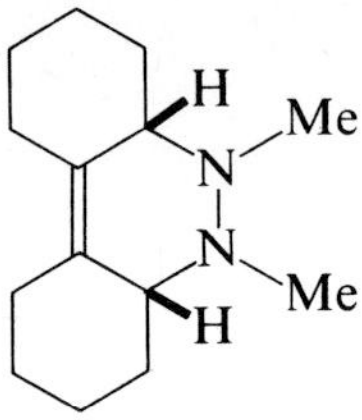   (5)

Two X-ray crystallographic investigations have been reported.[32b] In compound (4) which has both methyl groups equatorial in the solid, the CN—NC and MeN—NMe torsion angles are 65° and 64° respectively. The ring is thus considerably more puckered about the N—N bond than is cyclohexane. Compound (5) is observed to have an axial–equatorial conformation and the dihedral angles here are CN—NC, 67° and MeN—NMe, 72°. These values also indicate an increased puckering of the ring. The nitrogen atoms in the latter compound are found to be appreciably flatter ($\beta = 49°$, 48°) than in the former compound $\beta = 59°$ (Fig. 5.1).

The conformational route map for N,N'-dimethylhexahydropyridazine is shown in Fig. 5.2. There are three distinct types of conformation, diequatorial (ee), axial–equatorial (ae or ea) and diaxial (aa). These conformations are separated by four distinct types of process: two ring inversions ee $\rightleftharpoons$ aa and ae $\rightleftharpoons$ ea, and two nitrogen inversions ee $\rightleftharpoons$ ae or ea, and aa $\rightleftharpoons$ ea or ae. As we discussed earlier, the barriers to nitrogen inversion processes are raised by the adjacent nitrogen atom giving barriers of similar magnitude to those for ring inversion. The system therefore has three ground state energies and four transition state energies, and its complete elucidation has posed a considerable challenge.

Despite several false starts[33] application of a combination of variable temperature ^{13}C and ^{1}H NMR spectroscopy to the study of this system has elucidated the relative amounts of each conformation present, and the size of the barriers separating them.[34] The diequatorial conformation is 0.23 kcal mol^{-1} more stable than the axial–equatorial conformation. The diaxial conformation is considerably higher in energy since it is not observed in the low temperature spectra. The barriers to the four inversion processes have

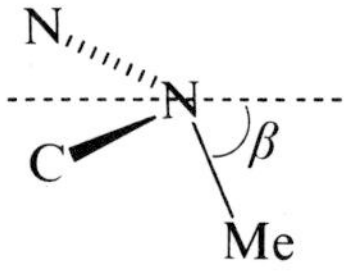

Fig. 5.1   Definition of angle to measure deformation of nitrogen atom.

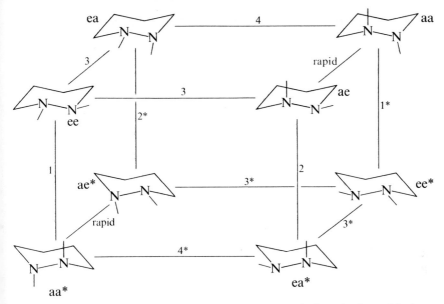

Fig. 5.2   Ring and nitrogen inversion in *N,N'*-dimethylhexahydropyridazine.

also been determined to be, in the nomenclature of Fig. 5.2: 1, 10.2; 2, 11.5; 3, 12.6 and 4, 7.5 kcal mol^{-1}.

The conformational free energy difference between the diequatorial and axial–equatorial conformations (0.23 kcal mol^{-1}) is very much less than in *N*-methylpiperidine (ca. 2.7 kcal mol^{-1}). Undoubtedly this difference arises from the different torsional potential about the N—N bond. In the hydrazine moiety in the ring the pairs of electrons on nitrogen have a distinct energetic bias towards a *gauche* (60°) arrangement.

Other methods of study that have been applied fruitfully to the study of this ring system are photoelectron spectroscopy,[35-38] and low-temperature cyclic voltammetry.[39,40]

Hexahydropyrimidine

The hexahydropyrimidine ring has proved less puzzling than the hexahydropyridazine because the β nitrogen atom increases the rate of nitrogen inversion relative to piperidine. No ambiguity arises therefore in low-temperature NMR spectra. The barriers to ring inversions are found to be ca 11 kcal mol^{-1},[41-43] whereas the barrier to nitrogen inversion is very much less than this.[1]

The most interesting conformational question in this series relates to the position of the conformational equilibrium on nitrogen. This problem has been solved by ^{1}H NMR,[44, 45] ^{13}C NMR[46] and dipole moment,[47] measurements. The N-methyl derivative (6) is found, from measurements of the N—H to C-2 methylene coupling constants, to have an axial N—H group.[44]

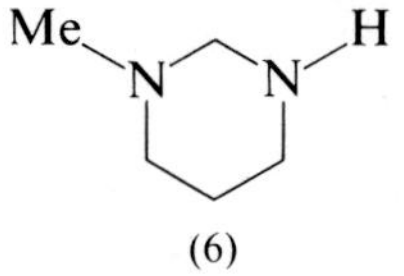

(6) 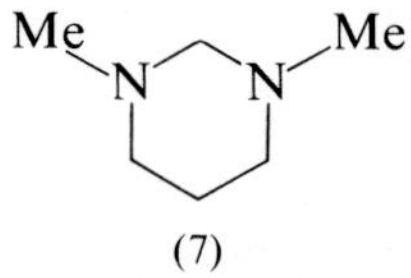   (7)

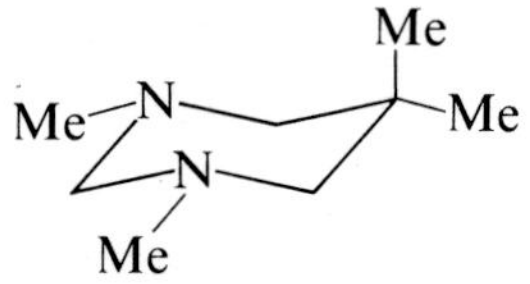

(8) 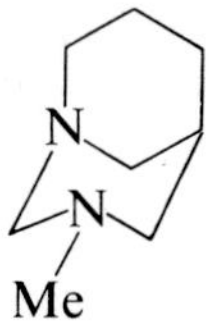   (9)

The N,N'-dimethyl derivative (7) was studied by comparing its spectral characteristics with the anancomeric models (8) and (9).[45] The diequatorial conformation is ca 0.5 kcal mol^{-1} more stable than either of the enantiomeric axial–equatorial conformations. However, with an entropy of mixing of Rln2 favouring the axial-equatorial conformation the apparent free energy difference is reduced to ca 0 kcal mol^{-1} and both conformations are nearly equal in population. Carbon-13 NMR results are in agreement with these results giving $\Delta G°$ 0.65 $\pm$ 0.05 kcal mol^{-1} and $\Delta G‡$ 7.0 $\pm$ 0.3 kcal mol^{-1}.[46] Values have also been obtained for the free energy differences of other alkyl groups on nitrogen.[45, 47] The lowering of the free energy difference in this system with respect to N-methylpiperidine is attributed to a combination of dipolar and steric factors. The "anomeric effect" will favour an axial–equatorial arrangement of N-substituents, which contains a lower dipole–dipole interaction. Replacement of an axial ring hydrogen atom by an electron pair on nitrogen further lowers the energy of an N-alkyl group due to smaller steric interactions.

An interesting series of changes is observed in the ^{1}H NMR spectra of the bishexahydropyrimidylmethanes (10), as the temperature is lowered.[48] At the low temperature limit, spectra from two sets of conformations, separated

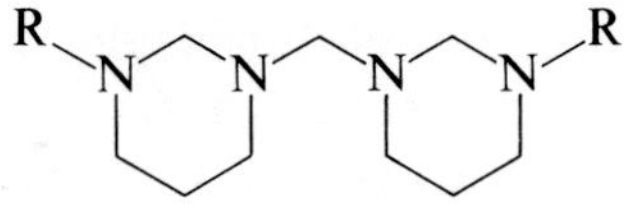

(10)

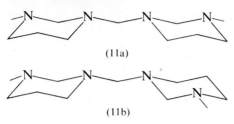

(11a)

(11b)

by ring inversion, are observed. The *meso* set (11a) shows an AB quartet for the bridging methylene group whilst the racemic set (11b) shows a singlet. Two AB quartets for the ring N—CH$_2$—N groups are observed.

Piperazine

The conformational properties of piperazine (12) and its derivatives are very closely related to those of the piperidines. Infrared studies in the N—H stretching first overtone region show that the N—H equatorial conformation of *N-t*-butylpiperazine is lower in enthalpy than the N—H axial conformation.[49] In *N,N'*-dimethylpiperazine the axial–equatorial free energy difference (2.96 ± 0.05 kcal mol^{-1}), measured by kinetically controlled protonation,[27] is very close to that found in *N*-methylpiperidine. Finally, the ring inversion barriers in the piperazines are almost identical to those in the piperidines (piperidine 10.4, vs piperazine 10.3, and *N*-methylpiperidine 11.9 vs *N*-methylpiperazine 11.5 kcal mol^{-1}).[50]

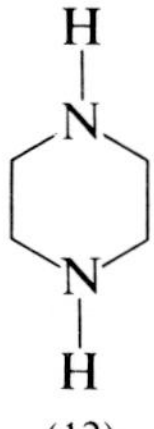

H
|
N

N
|
H

(12)

The similarities between the piperidine and the piperazine systems indicate the minimal effect of γ heterosubstitution on the conformational behaviour of heterocyclic six-membered rings. The anomalous free energy differences in *N*-chloropiperidine and *N,N'*-dichloropiperazine, however, remain to be explained.[29, 30]

Evidence has been presented from ^{13}C NMR spectra that whereas *trans*-1,4-dibenzoyl-2,5-dialkylpiperazines have a chair conformation their *cis* isomers exist in twist conformations, a result confirmed by X-ray crystallography.[51]

Hexahydro-1,3,5-Triazine

The effect of introducing a heteroatom β to nitrogen has been shown above to reduce the axial equatorial free energy difference of substituents on nitrogen substantially. In the hexahydro-1,3,5-triazine series (13) this effect is enhanced considerably due to the presence of a second β-nitrogen.[52,53] Thus the preferred conformation of the trimethyl derivative (13, R = Me) has one axial and two equatorial N-methyl groups (14a). This conformation is substantially more stable than the triequatorial conformation (14b) or

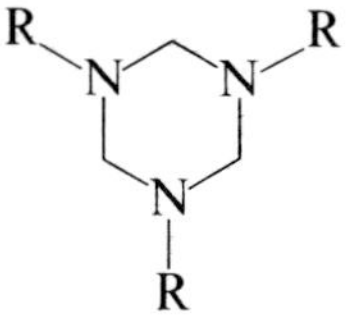

(13)

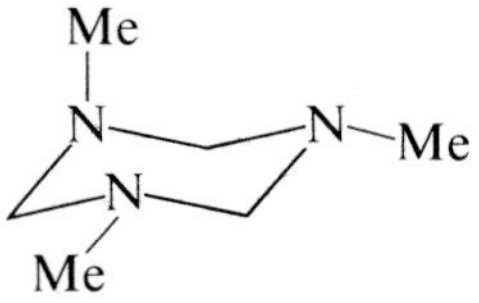

(14a)

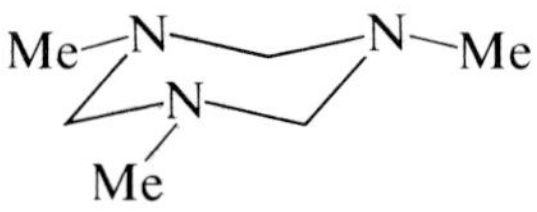

(14b)

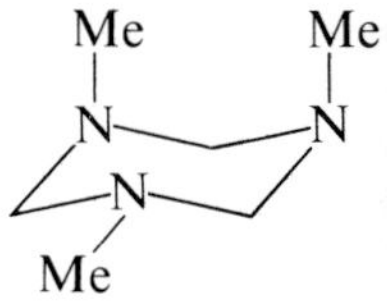

(14c)

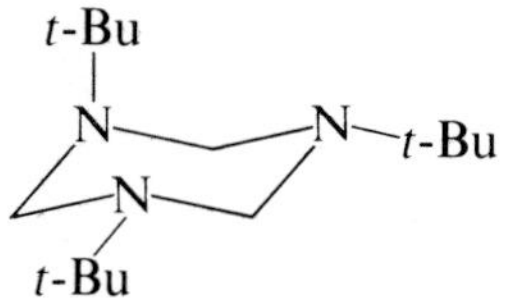

(15a)

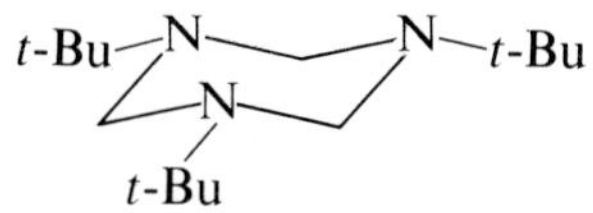

(15b)

the monoequatorial conformation (14c), neither of which can be detected in the ^{1}H NMR spectrum at -144° when both ring and nitrogen inversion are slow.[53] The preference of this ring system to have one N-substituent axial persists even in the tri-t-butyl derivative, where dipole moment measurements suggest that (15a) is ca 0.35 kcal mol^{-1} more stable than (15b).[52] Ring inversion and nitrogen inversion parameters have been measured for several derivatives.[53]

Hexahydro-1,2,4,5-Tetrazine

The literature has carried a dispute on the conformational equilibria in these compounds.[54,55] The commonly accepted view now[55] is that the

(16a)   (16b)   (16c)

tetramethyl derivative (16) exists as a mixture of the diaxial–diequatorial conformation (16a) (30%) and the monoaxial conformation (16b) (70%).

These results make sense in the context of what is now known about the conformational equilibria in the hexahydropyridazine and hexahydropyrimidine fragments from which this ring is composed. The diequatorial conformation of the corresponding pyridazine is only marginally more stable than the axial–equatorial (0.23 kcal mol^{-1}) and there is a distinct tendency in the hexahydropyrimidine ring for N-methyl groups to go axial. The combination of these tendencies gives rise to the observations, but does not explain the seeming absence of the centrosymmetric diaxial conformation (16c) advocated by Anderson and Roberts and by Nelsen and Hintz.[54]

Tetrahydro-1,2-Oxazine

The tetrahydro-1,2-oxazine system has been extensively studied by Riddell et al.[3, 56–61] and by Katritzky's group.[62, 63] The important heterocyclic feature of the ring is the N—O bond, a hydroxylamine fragment. As has been discussed above this fragment will affect torsional behaviour on the N—O side of the ring, and will slow down rates of nitrogen inversion. Both these expected effects are observed.

In an X-ray crystallographic study the torsion angle about the N—O bond is found to be 67° (cf. cyclohexane ca 55°)[59] whilst the alicyclic side of the ring retains "normal" dihedral angles. In the parent compound, tetrahydro-1,2-oxazine, infrared results show that the N—H group is almost exclusively equatorial.[62, 63] In the ^{1}H NMR spectra no trace can be found of an axial N-alkyl group in alkyl derivatives at temperatures where both nitrogen and ring inversion are expected to be slow.[57] There is thus a very strong tendency for N-substituents to be equatorial. Both this strong equatorial tendency and the puckering of the ring about the N—O bond are completely in accord with what is known about the shape and magnitude of the N—O bond torsional potential.

The oxygen atom adjacent to the nitrogen slows down the rate of nitrogen inversion,[1–3, 56] to such an extent that its rate becomes measurable by NMR

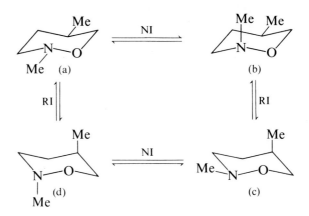

Fig. 5.3   Ring and nitrogen inversion in 2,5-dimethyltetrahydro-1,2-oxazine.

spectroscopy at around $0°$. Somewhat below that temperature (at ca $-40°$) conformations separated by nitrogen inversion barriers are observable. Thus, slowing of the nitrogen inversion in 2,5-dimethyltetrahydro-1,2-oxazine[61] (Fig. 5.3) separates (a) and (d) from (b) and (c). At low temperatures two sets of signals are observed which are attributed to (a) (major) and (c) (minor) since the axial N-methyls in (b) and (d) will considerably raise the energy of these species. The observed free energy difference is 1.36 ± 0.1 kcal mol^{-1}. Analogous experiments allowed the determination of the free energy differences of methyl groups on all four ring carbon atoms[60] (Fig. 5.4).

From the X-ray crystallographic study it had proved possible to estimate the distances between axial methyl groups and other atoms on the ring hindering them (Fig. 5.5). Methyl groups at C(4) and C(5) are seen to have almost identical distances to the hindering ring atoms, yet experimentally the C(5) methyl is found to experience smaller interactions. This presumably arises from the smaller van der Waals radius of oxygen compared to nitrogen. An analogous comparison of positions 3 and 6 again shows similar trans-annular distances but the methyl group at C(3), forced against oxygen when

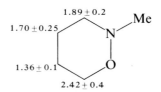

Fig. 5.4   Free energy differences (kcal mol^{-1}) of methyl groups at positions 3, 4, 5, and 6 in the tetrahydro-oxazine ring.

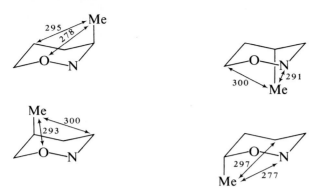

Fig. 5.5   Distances separating methyl carbon atoms from other ring atoms in tetrahydro-1,2-oxazines.

axial, experiences lower repulsions than the group at C(6) forced against nitrogen.

The relative sizes of oxygen and nitrogen atoms are thus important in the conformational equilibria in this ring system.

Tetrahydro-1,3-Oxazine

The tetrahydro-1,3-oxazine ring displays both of the expected consequences of introduction of a heteroatom β to nitrogen. The barrier to nitrogen inversion of an equatorial group in the N-methyl compound (7.6 kcal mol^{-1})* is lower than in N-methylpiperidine[4] (although the exact value of this parameter is still somewhat uncertain) and considerably lower than the barrier to ring inversion (10.0 kcal mol^{-1}).[3, 64]

Because of dipolar effects and reduced transannular interactions on axial N-alkyl groups, considerable amounts of axial-N-alkyl conformations are observed by NMR,[20, 44, 65–67] infrared, and dipole moment measurements.[49, 68] Booth and Lemieux[44] observed the low-temperature ^{1}H NMR spectrum of the parent compound (17), finding that the N—H proton had

(17)

* Reference 1 gives 6.8 kcal mol^{-1} but the authors have subsequently revised this upwards to 7.6 kcal mol^{-1} (A. R. Katritzky, personal communication).

couplings of 13.1 and 2.9 Hz with the C(2) methylene hydrogens. Clearly the predominant conformation has an axial N—H group. This conclusion is reinforced by infrared and dipole moment measurements from the Norwich group.[49, 68] N-Methyl derivatives have been shown to contain roughly equal amounts of axial and equatorial conformations,[20, 65–67] by comparing their NMR spectra with those of anancomeric models. The free energy difference for an N-methyl group is found to be 0.0 ± 0.35 kcal mol^{-1}.[20] In contrast it appears that in the N-protonated form this free energy difference is much higher (1.0 ± 0.3 kcal mol^{-1}).[20] This value is higher than in the free base because of removal of the dipolar interactions, but less than in cyclohexane because of the ring oxygen atom β to nitrogen removing a syn-axial hydrogen atom.

Tetrahydro-1,4-Oxazine (Morpholine)

The heteroatoms in the tetrahydro-1,4-oxazine ring (18) appear to be sufficiently far apart to exert a minimal influence on each other. The first overtone region of the N—H stretching frequency was studied by Baldock and Katritzky.[22] They found the enthalpy difference between the axial and equatorial N—H groups to be almost identical to that in piperidine, favouring the equatorial conformation by 0.63 kcal mol^{-1}. This work is further substantiated by the apparent predominance of lines due to the N—H equatorial conformation in the microwave spectrum.[69] In addition the ring inversion barrier of N-methylmorpholine (11.5 kcal mol^{-1}) is almost identical to that of N-methylpiperidine (11.9 kcal mol^{-1}).[50]

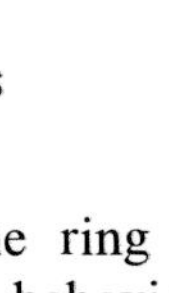

(18)

Tetrahydro-1,4,2-Dioxazines

The tetrahydro-1,4,2-dioxazine ring (19) has given some fascinating conformational results.[70–72] The behaviour of this dioxazine ring should be composed of aspects of the behaviour of the previously discussed 1,2- and 1,3-oxazine rings and gives an interesting insight into the way the conflicting conformational tendencies of these two rings may be fused together.

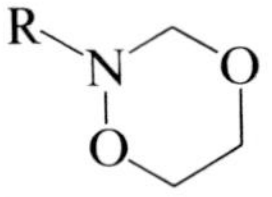

(19)

The NMR spectra of the N-methyl derivative (19, R = Me) show evidence of two coalescence phenomena arising from slowing of nitrogen inversion ($\Delta G^{\ddagger} 11.4 \pm 0.2 \, \text{kcal mol}^{-1}$) and ring inversion ($\Delta G^{\ddagger} 10.9 \pm 0.2 \, \text{kcal mol}^{-1}$). It is quite clear that the barrier to nitrogen inversion is intermediate between those in the 1,2-[56] and 1,3-oxazines,[1] and clearly demonstrates the barrier-lowering effect of the β-oxygen atom (or the barrier-raising effect of the α oxygen) on this inversion of an equatorial N-methyl group.

The conformational free energy difference of the N-methyl group is found to be ca 1 kcal mol^{-1}. Again this is between the values for the 1,2-oxazines (where it is too large to be measured) and the 1,3-oxazines (ca 0 kcal mol^{-1}). The replacement of the syn-axial C—H bond by electrons on oxygen and the favourable anomeric interaction are responsible for this value.

Tetrahydro-Oxadiazine

Tetrahydro-1,2,4-oxadiazine

Just as the conformational properties of the 1,2,4-dioxazines described above reflect a combination of the properties of the tetrahydro-1,2-oxazine and 1,3-oxazine rings, so the tetrahydro-1,2,4-oxadiazine ring (20) combines the conformational tendencies of the tetrahydro-1,2-oxazine and hexahydropyrimidine rings.[73, 74] In the 2,4-dimethyl derivative (20, R = Me) a detailed conformational analysis has been possible by a combined use of ^{1}H and ^{13}C NMR spectra.[74] The equatorial $N(4)$-methyl inversion barrier is 7.4–7.9 kcal mol^{-1} with a free energy difference of 0.6–0.9 kcal mol^{-1}. These values are very similar to those found in the hexahydropyrimidine ring (see above). The equatorial $N(2)$-methyl inversion barrier is ca 12.7–13.1 kcal mol^{-1}. This is higher than in the 1,2,4-dioxazines but lower than in the 1,2-oxazine series, reflecting the fact that the additional β nitrogen atom lowers the N-inversion barrier, but by a smaller amount than oxygen. The barrier to ring inversion is ca 12.7 kcal mol^{-1}.

(20)

Tetrahydro-1,2,5-oxadiazine

In this series of oxadiazines (21) the "reference compounds" are the tetrahydro-1,2- and 1,3-oxazines. There is a kinetic process observed in the ^{1}H NMR spectrum with activation parameters so similar to those for 2-methyltetrahydro-1,2-oxazine as to strongly suggest $N(2)$ inversion as a common origin.[75, 76] The γ effect of nitrogen is therefore negligible.

For substituted derivatives, some interesting observations are recorded.[77] On freezing out of $N(2)$ inversion the 2,4,5-trimethyl derivative (22) shows peaks associated with a major conformational set (23 and 24) and a minor conformation (25). In the major set with the 4-methyl group equatorial the 5-N-methyl group equally populates the axial and equatorial positions. This

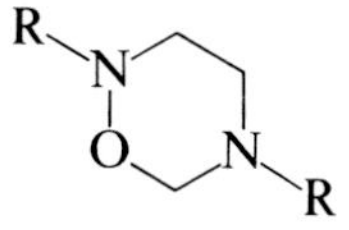

(21) 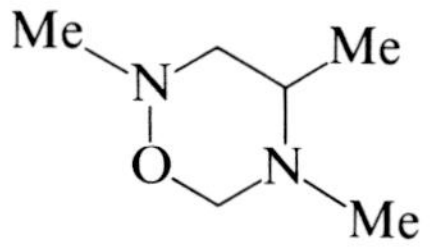 (22)

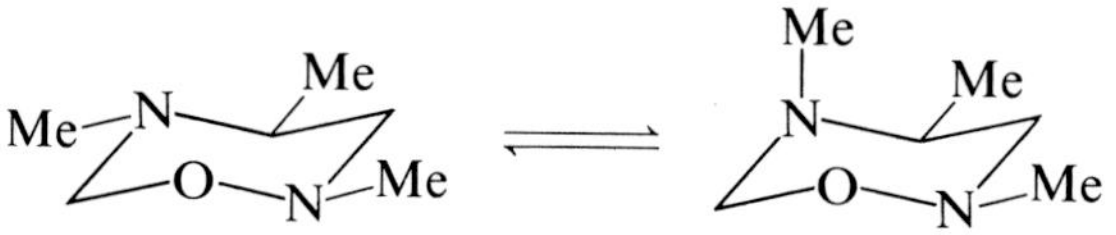

(23) (24) 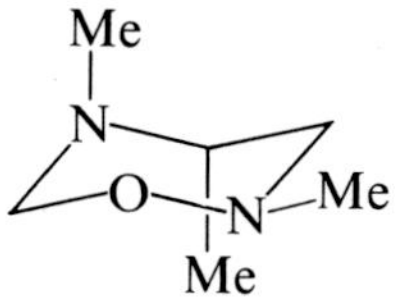 (25)

is in complete accord with the behaviour of the 1,3-oxazine moiety if the gauche-butane interactions about the 4–5 bond are roughly equal for axial and equatorial N-methyl. The minor conformation (25) has its 4-methyl and 5-methyl groups axial. Here the N-methyl group goes axial to avoid the gauche-butane interaction when it is equatorial. It is able to do this quite readily as it forms part of a 1,3-oxazine system.

Free energy differences for groups on the tetrahydro-1,2,5-oxadiazine ring are: 4-methyl, 1.2 ± 0.2; 6-p-nitrophenyl, 1.4 ± 0.2 kcal mol^{-1}.

Tetrahydro-1,3,4-oxadiazine

The model "reference compounds" for this system are hexahydropyridazine and tetrahydro-1,3-oxazine. It is found that the preferred conformation of the N,N'-dimethyl derivative is 3-axial-4-equatorial (26).[78, 79]

This conformation represents by far the best compromise between the conformational requirements of both contributory systems.[78]

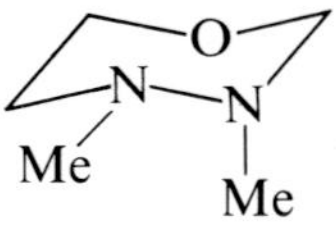

(26)

Nitrogen-Containing Fused Rings

The two most important nitrogen-containing fused rings are quino-lizidine (27) and decahydroquinoline (28).[80] In quinolizidine there are two possible conformations arising from inversion at nitrogen. That with the *trans*-fused rings (29a) is considerably more stable (ca 4.5 kcal mol^{-1}) than the *cis*-fused conformation (29b).[81,82] This is considerably greater than in the carbocyclic analogue decalin (ca 2.7 kcal mol^{-1}), but is not now un-expected in view of the modern value for the conformational free energy difference in *N*-methylpiperidine (ca 2.7 kcal mol^{-1}), which is considerably greater than for methylcyclohexane (ca 1.7 kcal mol^{-1}) (see above).

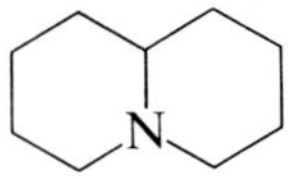

(27) 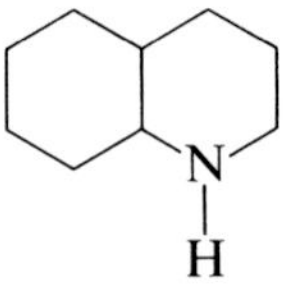   (28)

(29a) 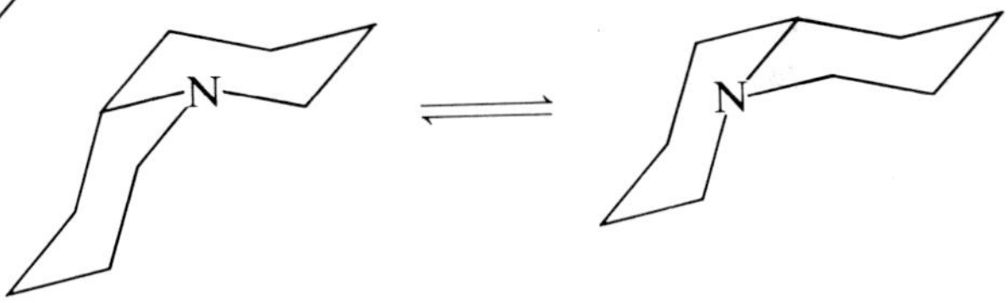   (29b)

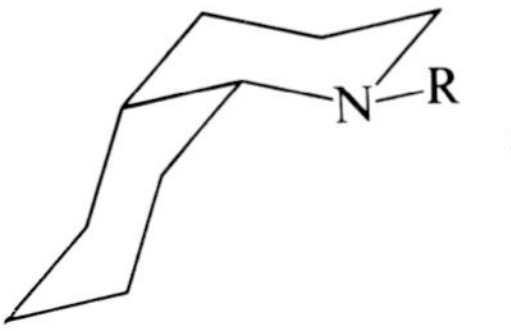

(30a) 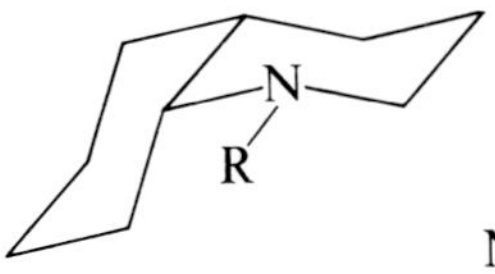   (30b) 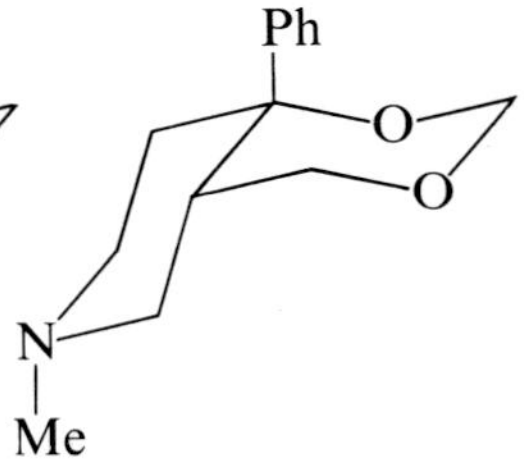   (31)

The conformational analysis of *trans*-decahydroquinoline,[83–85] *cis*-decahydroquinoline,[86,87] and of heterosubstituted derivatives of quinolizidine[80] and decahydroquinoline[88] follow the patterns outlined above for monocyclic systems. With *cis* ring fusion small heteroatoms such as O and N prefer *endo* positions as in *cis*-decahydroquinoline in which (30a) is preferred to (30b) or in the system (31) where the maximum number of *endo* oxygen atoms is found.[39]

References

1.   I. J. Ferguson, A. R. Katritzky and D. M. Read. (1975). *J.C.S. Chem. Comm.* 255.
2.   F. G. Riddell and H. Labaziewicz. (1975). *J.C.S. Chem. Comm.* 766.
3.   F. G. Riddell and J. E. Anderson. (1977). *J.C.S. Perkin II* 588.
4.   A. R. Katritzky, R. Patel and F. G. Riddell. (1979). *J.C.S. Chem. Commun.* 674.
5.   See for example, J. A. Pople and D. L. Beveridge. (1970). "Approximate Molecular Orbital Theory", 119 ff. McGraw-Hill, New York.
6.   J. J. Guy and T. A. Hamor. (1974). *J.C.S. Perkin II* 101.
7.   P. W. Codding and N. M. G. James. (1974). *Acta Cryst* **B30**, 240.
8a.   T. V. Tillack, R. C. Seccombe, C. H. L. Kennard and P. W. T. Oh. (1974). *Rec. Trav. Chim.* **93**, 164.
8b.   H. van Koningsveld. (1970). *Rec. Trav. Chim.* **89**, 375.
9.   See for example, C. Altona and M. Sundaralingham. (1970). *Tetrahedron* **26**, 925.
10.   For a detailed review of this controversy see, I. D. Blackburne, A. R. Katritzky and Y. Takeuchi. (1975). *Acc. Chem. Res.* **8**, 300.
11.   M. J. Aroney and R. J. W. Le Fevre. (1958). *J. Chem. Soc.* 3002.
12.   D. H. R. Barton and R. C. Cookson. (1956). *Quart. Rev.* **10**, 44.
13.   N. L. Allinger, J. G. D. Carpenter and F. M. Karkowski. (1964). *Tetrahedron Lett.* 3345; (1965). *J. Am. Chem. Soc.* **87**, 1232.
14.   R. J. Bishop, L. E. Sutton, D. Dineen, R. A. Y. Jones and A. R. Katritzky. (1964). *Proc. Chem. Soc.* 257.
15.   R. A. Y. Jones, A. R. Katritzky, A. C. Richards and R. J. Wyatt. (1970). *J. Chem. Soc.* (*B*) 122; R. J. Bishop, L. E. Sutton, D. Dineen, R. A. Y. Jones, A. R. Katritzky and R. J. Wyatt, (1967). *J. Chem. Soc.* (*B*) 493.
16.   M. J. Aroney, C. Y. Chen, R. J. W. Le Fevre and J. D. Saxby. (1964). *J. Chem. Soc.* 4269.
17.   J. B. Lambert and R. G. Keske. (1966). *J. Am. Chem. Soc.* **88**, 620.
18.   J. B. Lambert, R. G. Keske and D. K. Weary. (1967). *J. Am. Chem. Soc.* **89**, 5921.
19.   M. J. T. Robinson. (1968). *Tetrahedron Lett.* 1153.
20.   F. G. Riddell and J. M. Lehn. (1968). *J. Chem. Soc.* (*B*) 1224.
21.   P. J. Buckley, C. C. Costain and J. E. Parkin. (1968). *Chem. Comm.* 668. (Although the final paper has not yet appeared there is no reason to doubt the definitive nature of the results in this preliminary communication.)
22.   R. W. Baldock and A. R. Katritzky. (1968). *Tetrahedron Lett.* 1159; (1968). *J. Chem. Soc.* (*B*) 1470.
23.   F. A. L. Anet and I. Yavari. (1977). *J. Am. Chem. Soc.* **99**, 2794. See also F. W. Vierhapper and E. L. Eliel. (1979). *J. Org. Chem.* **44**, 1081.

24.   See F. G. Riddell. (1974). *In* "Internal Rotation in Molecules", Chapter 2, (Ed. Orville Thomas). Wiley, London.
25a. P. J. Crowley, M. J. T. Robinson and M. G. Ward. (1974). *J.C.S. Chem. Comm.* 825; (1977). *Tetrahedron.* **33**, 915.
25b. M. J. T. Robinson. (1975). *J.C.S. Chem. Comm.* 844.
25c. E. L. Eliel, D. Kandasamy, C-y. Yen and K. D. Hargrave, personal communication.
26.   D. C. Appleton, J. McKenna, J. M. McKenna, L. B. Simms and A. R. Walley. (1976). *J. Am. Chem. Soc.* **98**, 292.
27.   F. A. L. Anet and I. Yavari. (1976). *Tetrahedron Lett.* 2093.
28.   F. A. L. Anet, I. Yavari, I. J. Ferguson, A. R. Katritzky, M. Moreno-Mañas and M. J. T. Robinson. (1976). *Chem. Comm.* 399.
29a. K. W. Baldry and M. J. T. Robinson. (1975). *Tetrahedron* **31**, 2621.
29b. F. A. L. Anet and I. Yavari. (1977). *Tetrahedron Lett.* 3207.
30.   F. A. L. Anet and I. Yavari. (1978). *J.C.S. Chem. Comm.* 58.
31.   H. Ripperger. (1977). *Z. Chem.* **17**, 177.
32a. S. F. Nelsen. (1978). *Acc. Chem. Res.* **11**, 14.
32b. S. F. Nelsen, W. C. Hollinsed and J. C. Calabrese. (1977). *J. Am. Chem. Soc.* **99**, 4461.
33a. J. E. Anderson. (1969). *J. Am. Chem. Soc.* **91**, 6374.
33b. R. A. Y. Jones, A. R. Katritzky, D. L. Ostercamp, K. A. F. Record and A. C. Richards. (1971). *Chem. Comm.* 644.
33c. R. A. Y. Jones, A. R. Katritzky, K. A. F. Record and R. Scattergood. (1974). *J.C.S. Perkin II* 406.
34.   S. F. Nelsen and G. R. Weisman. (1976). *J. Am. Chem. Soc.* **98**, 5269.
35.   S. F. Nelsen and J. M. Buschek. *J. Am. Chem. Soc.* (1973). **95**, 2011; (1974). **96**, 2392, 6982, 6987.
36.   S. F. Nelsen, J. M. Buschek and P. J. Hinty. (1974). *J. Am. Chem. Soc.* **96**, 2013.
37.   P. Rademacher. *Angew Chem.* (1973). **85**, 410; *Tetrahedron Lett.* (1974). 83; *Chem. Ber.* (1975). **108**, 1548.
38.   P. Rademacher and H. Koopman. (1975). *Chem. Ber.* **108**, 1557.
39.   S. F. Nelsen, L. Echegoyen and D. H. Evans. (1975). *J. Am. Chem. Soc.* **97**, 3530.
40.   S. F. Nelsen, L. Echegoyen, E. L. Clennan, D. H. Evans and D. A. Corrigan. (1977). *J. Am. Chem. Soc.* **99**, 1130.
41.   F. G. Riddell and J. M. Lehn. (1966). *Chem. Comm.* 375.
42.   F. G. Riddell. (1967). *J. Chem. Soc.* (*B*) 560.
43.   R. F. Farmer and J. Hamer. (1968). *Tetrahedron* **24**, 829.
44.   H. Booth and R. U. Lemieux. (1971). *Can. J. Chem.* **49**, 777.
45a. F. G. Riddell and D. A. R. Williams. (1971). *Tetrahedron Lett.* 2073.
45b. E. L. Eliel, L. D. Kopp, J. E. Dennis and S. A. Evans, Jr. (1971). *Tetrahedron Lett.* 3409.
46.   A. R. Katritzky, V. J. Baker, I. J. Ferguson and R. C. Patel. (1979). *J.C.S. Perkin II* 143.
47.   R. A. Y. Jones, A. R. Katritzky and M. Snarey. (1970). *J. Chem. Soc.* (*B*) 131.
48.   F. G. Riddell. (1971). *J. Chem. Soc.* (*B*) 1028.
49.   M. J. Cook. R. A. Y. Jones, A. R. Katritzky, M. Moreno-Mañas, A. C. Richards, A. J. Sparrow and D. L. Trepanier. (1973). *J.C.S. Perkin II* 325.
50.   R. K. Harris and R. A. Spragg. (1968). *J. Chem. Soc.* (*B*) 684.
51.   S. Tsuboyama, K. Tsuboyama, J. Uzawa, M. Nakamura, K. Kobayashi and T. Sakurai. (1977). *Tetrahedron Lett.* 2895.

52a. R. A. Y. Jones, A. R. Katritzky and M. Snarey. (1970). *J. Chem. Soc.* (*B*) 135.
52b. R. P. Duke, R. A. Y. Jones, A. R. Katritzky, R. Scattergood and F. G. Riddell. (1973). *J.C.S. Perkin II* 2109.
52c. V. J. Baker, I. J. Ferguson, A. R. Katritzky, R. C. Patel and S. Rahimi-Rastgoo. (1978). *J.C.S. Perkin II* 377.
53a. C. H. Bushweller, M. Z. Lourandos and J. A. Brunelle. (1974). *J. Am. Chem. Soc.* **96**, 1591.
53b. J. M. Lehn, F. G. Riddell, B. J. Price and I. O. Sutherland. (1967). *J. Chem. Soc.* (*B*) 387.
53c. H. S. Gutowsky and P. A. Temussi. (1967). *J. Am. Chem. Soc.* **89**, 4358.
54a. J. E. Anderson and J. D. Roberts. (1968). *J. Am. Chem. Soc.* **90**, 4186.
54b. S. F. Nelsen and P. J. Hintz. (1972). *J. Am. Chem. Soc.* **94**, 3138.
55. R. A. Y. Jones, A. R. Katritzky, A. R. Martin, D. L. Ostercamp, A. C. Richards and J. M. Sullivan. *J. Am. Chem. Soc.* (1974). **96**, 576; *J.C.S. Perkin II* (1974). 948.
56a. F. G. Riddell, J. M. Lehn and J. Wagner. (1968). *Chem. Comm.* 1403.
56b. F. G. Riddell, E. S. Turner and A. Boyd. (1979). *Tetrahedron* **35**, 259.
57. F. G. Riddell, D. A. R. Williams, C. Hootele and N. Reid. (1970). *J. Chem. Soc.* (*B*) 1739.
58. F. G. Riddell and D. A. R. Williams. (1974). *Tetrahedron* **30**, 1083.
59. F. G. Riddell, P. Murray-Rust and J. Murray-Rust. (1974). *Tetrahedron* **30**, 1087.
60. F. G. Riddell and D. A. R. Williams. (1974). *Tetrahedron* **30**, 1097.
61. F. G. Riddell. (1975). *Tetrahedron* **31**, 523.
62. R. A. Y. Jones, A. R. Katritzky, A. C. Richards, S. Saba, A. J. Sparrow and D. L. Trepanier. (1972). *J.C.S. Chem. Comm.* 673.
63. R. A. Y. Jones, A. R. Katritzky, S. Saba and A. J. Sparrow. (1974). *J.C.S. Perkin II* 1554.
64. J. M. Lehn, P. Linscheid and F. G. Riddell. (1968). *Bull. Soc. Chim. France* 1172.
65. Y. Allingham, R. C. Cookson, T. A. Crabb and S. Vary. (1968). *Tetrahedron* **24**, 4625.
66. T. A. Crabb and R. F. Newton. (1968). *Tetrahedron* **24**, 4423.
67. T. A. Crabb and S. I. Judd. (1970). *Org. Mag. Res.* **2**, 317.
68. R. A. Y. Jones, A. R. Katritzky and D. L. Trepanier. (1971). *J. Chem. Soc.* (*B*) 1300.
69. J. J. Sloan and R. Kewley. (1969). *Canad. J. Chem.* **47**, 3453.
70. R. A. Y. Jones, A. R. Katritzky, A. R. Martin and S. Saba. (1973). *J.C.S. Chem. Comm.* 908.
71. R. A. Y. Jones, A. R. Katritzky, A. R. Martin and S. Saba. (1974). *J.C.S. Perkin II* 1561.
72. F. G. Riddell, M. H. Berry and E. S. Turner. (1978). *Tetrahedron* **34**, 1415.
73. F. G. Riddell and E. S. Turner. (1978). *Heterocycles* **9**, 267.
74. F. G. Riddell, E. S. Turner, A. R. Katritzky, R. C. Patel and F. M. S. Brito-Palma. (1979). *Tetrahedron* **35**, 1391.
75. A. R. Katritzky and R. C. Patel. (1978). *Heterocycles* **9**, 263.
76. F. G. Riddell and E. S. Turner. (1978). *J. Chem. Res.* (*S*) 476.
77. F. G. Riddell and E. S. Turner. (1979). *Tetrahedron* **35**, 1131.
78. F. G. Riddell and A. J. Kidd. (1977). *J.C.S. Perkin II* 1816.
79. I. J. Ferguson, A. R. Katritzky and D. M. Read. (1976). *J.C.S. Perkin II* 1861.
80. For a comprehensive review see, T. A. Crabb, R. F. Newton and D. Jackson. (1971). *Chem. Rev.* **71**, 109.
81. H. S. Aaron. (1965). *Chem. Ind.* 1338.

82. C. D. Johnson, R. A. Y. Jones, A. R. Katritzky, C. R. Palmer, K. Schofield and R. J. Wells. (1965). *J. Chem. Soc.* 6797.
83. H. S. Aaron and C. P. Ferguson. (1976). *J. Am. Chem. Soc.* **98**, 7013.
84. E. L. Eliel and F. W. Vierhapper. (1974). *J. Am. Chem. Soc.* **96**, 2257.
85. E. L. Eliel and F. W. Vierhapper. (1976). *J. Org. Chem.* **41**, 199.
86. E. L. Eliel and F. W. Vierhapper. (1977). *J. Org. Chem.* **42**, 51.
87. H. Booth and D. V. Griffiths. (1973). *J.C.S. Perkin II* 842.
88. See for example, R. Cahill and T. A. Crabb. (1972). *Org. Mag. Res.* **4**, 283.
89. A. F. Casey, A. B. Simmonds and D. Staniforth. (1972). *J. Org. Chem.* **37**, 3189; G. Swaelens and M. Anteunis. (1970). *Tetrahedron Lett.* 561.

6

Six-Membered Rings Containing
Sulphur and Phosphorus

Sulphur-Containing Rings

General considerations

There are two effects that differentiate six-membered sulphur-containing rings from those compounds so far discussed in this book. The C—S bond is longer than the C—C, C—O, or C—N bonds we have until now been considering. A typical C—S bond length is ca 181 pm. Since sulphur is a second-row element its van der Waals radius is larger than those of oxygen or nitrogen but probably smaller than a methylene group (Table 2.1). More "conformational anomalies" are thrown up in sulphur compounds than in any others so far investigated, and the clarification of these points still presents a challenge. A review of the conformational analysis of sulphur-containing rings has been presented by Zefirov.[1]

Thian

Most of the published work on the conformational analysis of thians is concerned with *S*-substituted derivatives. Nevertheless several experimental and theoretical studies have been published on non-*S*-substituted derivatives.[2-4] The ring inversion barrier has been measured,[2,3] and is found to be 11.7 kcal mol^{-1}. Allinger and Hickey have developed a molecular mechanics force field incorporating thioethers.[4] Their calculations have given estimates of values for the free energy differences of methyl groups at various positions in the thian ring. These were calculated to be: 2, 0.99; 3, 1.10 and 4, 1.59 kcal mol^{-1}, whilst the twist conformation was calculated to be 4.03 kcal mol^{-1} above the chair. Experimental values measured by Willer and Eliel[5] are: 2 Me, 1.42; 3 Me, 1.40; 4 Me, 1.80 kcal mol^{-1}. These values are 0.3–0.4 kcal mol^{-1} higher than Allinger's calculations. The observed free energy difference at the 4-position is similar to that in methyl-cyclohexane, because the axial and equatorial environments are similar in both systems. For the 3-methyl group one *syn*-axial C—H bond has been replaced by an electron cloud on sulphur reducing the transannular axial interactions. For the 2-position the axial substituent may lean out of the ring lowering the free energy difference.

The crystal and molecular structure of *S*-methylthianium iodide (1)[6] gives the best available evidence for the conformation and molecular geometry of the thian ring. The main features of interest are as follows. The ring is more puckered than cyclohexane. This is shown by the dihedral angles between the C(2), C(3), C(5), C(6), plane and respectively the C(2), S, C(6) and the C3, C(4), C(5) planes being 126° and 123.7° (Fig. 6.1). In cyclohexane this parameter is 132.1°. Also the C(2)–C(6) distance is greater than the C(3)–C(5) distance, despite the C—S—C angle (100°) being smaller than tetrahedral because the C—S bonds (180 pm) are longer than the C—C bonds (ca 152 pm). Finally, and of relevance to Allinger's calculations,[4] and Eliel's experiment results[5], the axial hydrogens on C(2) and C(6) are found to be farther apart (271 pm) than those on C(3) and C(5) (244 pm).

(1)   (2a)   (2b)

The conformational free energy difference of an *S*-methyl group in thianium salts has been measured by Eliel *et al.*[7a] who equilibrated 4-*t*-butyl-*S*-methylthianium perchlorate (2) by heating in chloroform.* At 100° the equilibrium form (2a) is more stable by 0.275 kcal mol^{-1}. An X-ray crystal-lographic investigation of the isomers showed that there is considerable

* This equilibration reaction had previously been described by Katritzky *et al.*[7b]

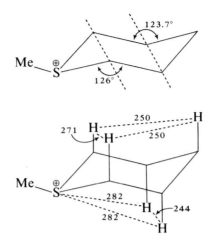

Fig. 6.1   Partial molecular geometry of the *S*-methylthianium ion with distances in pm (Ref. 6).

distortion of the axial isomer (2b) around the C(2), S, C(6) end of the ring to cope with the axial methyl group. The C(2, 6) bond angles expand from 107° in the equatorial isomer to 115°, whilst the ring dihedral angles along the S–C(2) and C(2)–C(3) bonds at 64° and 69° respectively with the equatorial methyl group decrease to 46° and 59° when the methyl is axial. It is not clear why these gross alterations in ring geometry should be associated with such a small free energy difference unless the molecules deform very readily.

Hydrogen-1 and ^{13}C NMR spectra have also been used to obtain conformational information on thianium ions.[8, 9] Although the hydrogen atoms at C(2) and (6) in thianium ions are somewhat acidic there seems to be little selectivity between axial and equatorial hydrogens in their exchange reaction with deuterium.[10]

Coupling constant measurements on the S—H resonance indicate that *S*-protonated thian exists with the S—H axial.[2] A similar preference for oxygen to be axial in sulphoxides (3) has been known for some time.[11] It has been quantitatively found to be 0.175 ± 30 kcal mol^{-1} by Lambert and Keske[12] using ^{1}H NMR. This result has been confirmed by Buchanan using ^{13}C NMR.[13] The activation energy for ring inversion is ca 14 kcal

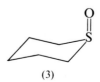

(3)

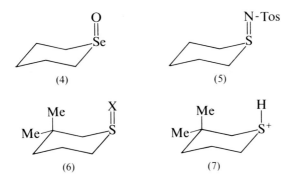

(4)   (5)   (6)   (7)

mol^{-1}.[12] A similar, although slightly greater axial preference, is found in selenan-1-oxide (4).[14]

The preference for axial substitution persists when =O is replaced by =N-Tos (5), but is inhibited by a 3-axial methyl group (6) although the protonated form of 3,3-dimethylthian has the S—H axial (7).[15]

It is generally accepted that *attractive* van der Waals interactions are operating in (3), (4), and (5) between the axial group and the *syn*-axial hydrogens. The "long" C—S bonds are placing these groups in the feebly attractive portion of the interaction potential. However, in (6) the axial methyl, being much larger than hydrogen causes a repulsive interaction. This is not felt by the axial S—H group which remains axial whether there is a 3-axial methyl or not.

1,2-Dithian

Very little work has been reported on this system. X-Ray diffraction studies of the *trans*-3,6-dicarboxyllic acids of 1,2-dithian (8) and 1,2-diselenan have been reported.[16] The tendency of the C—S—S—C dihedral angle to open to its preferred angle of ca 90° is seen in a marginal increase in (8) where it is 60° over the 55° value found in cyclohexane. 1,2-Dithian-1-oxide (9) exists in the O-axial conformation, whilst the dioxide (10) exhibits a rapid ring inversion in the NMR spectrum even as low as $-90°$.[17]

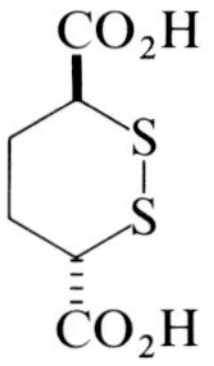

CO$_2$H

CO$_2$H

(8) 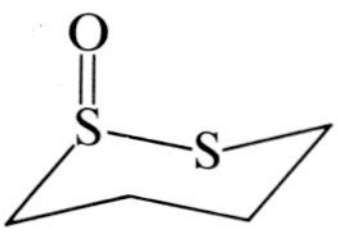   (9) 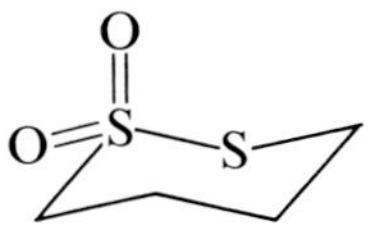   (10)

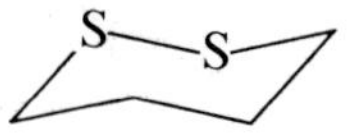

(11) 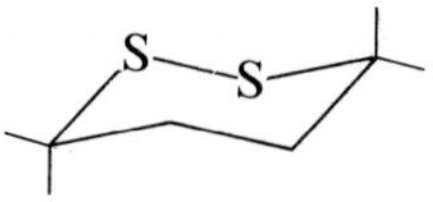   (12)

Variable-temperature ^{1}H NMR studies show a free energy of activation for ring inversion in (11) and (12) to be 11.6 and 13.8 kcal mol^{-1} respectively.[18]

1,3-Dithian

As with the 1,3-dioxans, the 1,3-dithian system has proved a very popular vehicle for conformational studies. The reasons are similar: their ease of synthesis with a wide variety of substituents, their readily interpretable NMR spectra and their easy equilibration reactions.

The structure of the basic 1,3-dithian ring has been found from an X-ray diffraction study of 2-phenyl-1,3-dithian (13).[18] The dihedral angles in the ring about the various bonds are similar to or slightly greater than those in cyclohexane, showing the ring to be marginally more puckered. A similar result is obtained from R-value measurements.[19] X-Ray crystallographic studies of the derived sulphoxides (14–16) show that increasing sulphoxide substitution increases the puckering of the ring.[20] For example, the ring torsion angles around C(5) in the disulphoxide (16) are 72.5°, dramatically larger than in cyclohexane. A similar result is obtained for the series (17–19).[21]

As with the 1,3-dioxan series the 1,3-dithians are amenable to acid-catalysed equilibration reactions. Eliel and Hutchins measured the values

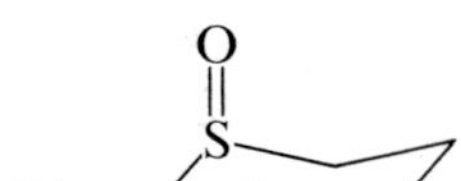

(13) 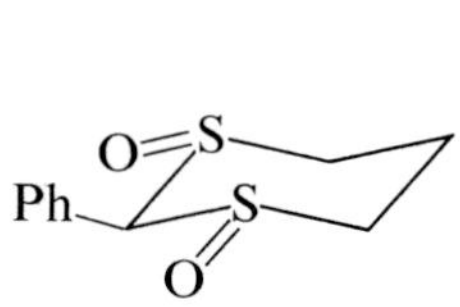   (14)

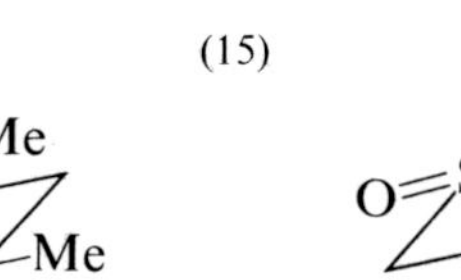

(15)   (16)

(17)   (18)   (19)

TABLE 6.1   Conformational free energy differences of alkyl substituents
in the 1,3-dithian ring (kcal mol^{-1}) (from Ref. 22).

2 Me 1.77	2 Et 1.54	2 i-Pr 1.95	2 t-Bu > 2.7
4 Me 1.69	—	—	—
5 Me 1.05	5 Et 0.77	5 i-Pr 0.8	—

recorded in Table 6.1.[22] As has been noted in the 1,3-dioxan series and elsewhere it is slightly easier to put an ethyl group axial than a methyl. A few years later Pihlaja pointed out that the chair–twist free energy difference in 1,3-dithian was sufficiently low that twist conformations might contribute to the conformational composition of the least stable isomers reported by Eliel, and he therefore revised the conformational energy parameters slightly to take this into account.[23] He derived the following values for chair → twist change in 1,3-dithian: $\Delta H = 4.04 \pm 0.29$ kcal mol^{-1}; $\Delta S = 4.55 \pm 0.48$ J mol^{-1} K^{-1}, and confirmed these values in a subsequent paper.[24]

The chair–twist energy difference in 1,3-dithian is therefore smaller than in cyclohexane, which in turn is found to be less than in 1,3-dioxan. It is believed that this arises from the relative lengths of C—O, C—C, and C—S bonds making twist conformations progressively less compact along the series dioxan–cyclohexane–dithian, and so reducing the energy difference along the series.

Two ^{13}C NMR studies of 1,3-dithian[25,26] and one of 1,3-dithian oxides[27] have appeared. In all three papers it was shown that because of the low chair–twist energy difference in 1,3-dithian, twist conformations played a significant role in compounds with only a moderate degree of steric hindrance.

Two groups have shown the axial preference of a 5-hydroxyl group in a 1,3-dithian ring although the values of the free energy difference differed: 0.8 kcal mol^{-1} [28] and 0.5 kcal mol^{-1}.[29]

1,3-Dithian-2-carboxylic acids show an anomeric effect in the axial preference of a 2-carboxyl group.[30] This effect is also seen in the 1,3,5-trithian series and is in the opposite direction to that in the 1,3-dioxan series. Although the authors of the paper discuss this anomaly[30] no satisfactory theoretical explanation for this discrepancy has been generally agreed.

Low temperature ^{1}H spectra at 270 MHz on 1,3-dithian-1-oxide show two conformations in the ratio 84:16 at $-81.5°$ ($\Delta G = 0.63$ kcal mol^{-1}), the major conformation having the oxide equatorial (20).[31] This result

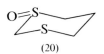

(20)

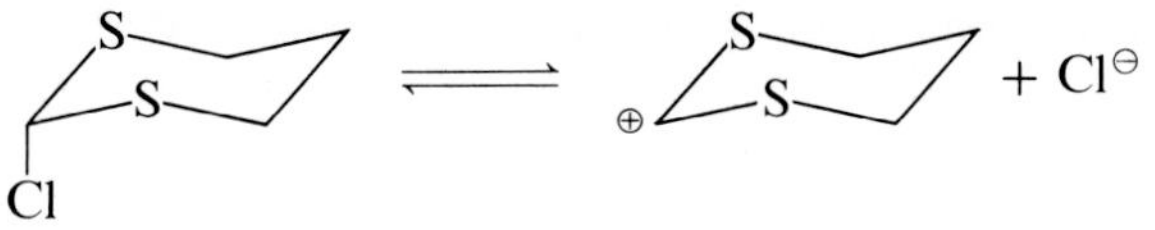

Fig. 6.2   Postulated ionisation–recombination mechanism in 2-chloro-1,3-dithian.

confirmed other earlier reported results.[32,33] Interestingly, 1,3-oxathian-3-oxide shows the opposite tendency.[33] In this case the axial conformation (21) is preferred by 0.57 kcal mol^{-1} at $-98°$. This effect is exactly the opposite of what might be expected on steric and dipolar grounds, and once again a satisfactory explanation is awaited.

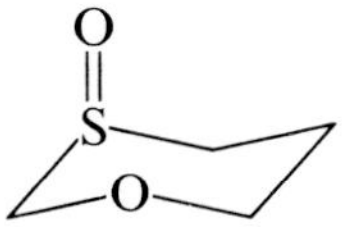

(21)

It has been suggested that the temperature variation of the ^{1}H NMR spectrum of 2-chloro-1,3-dithian is due to an ionization–recombination reaction (Fig. 6.2) rather than a ring inversion process.[34]

1,4-Dithian

Although a considerable amount of structural information from X-ray diffraction studies is available concerning the 1,4-dithian ring (22),[35–43] very little work concerned with conformational equilibria in derivatives has been carried out. Amongst the structures that have been solved are the parent compound[35] and its adducts with iodine,[36] iodoform,[37] antimony trioxide[38] and diiodoacetylene.[39] The α- (or trans)-dioxide (23) crystallizes with both oxygen atoms axial[40] whilst the β-dioxide (24) has one axial and one equatorial sulphoxide group.[41] Trans-2,3-Dichloro-1,4-dithian has both halogens axial for the favourable anomeric effect, and as in the 1,4-dioxan series the chlorines lean out from the ring to minimize gauche repulsive interactions with the other sulphur atom.[42] Trans-2,5-Dibromo-1,4-dithian has both bromines axial.[43]

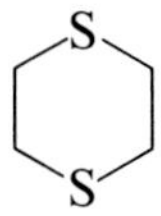

(22) 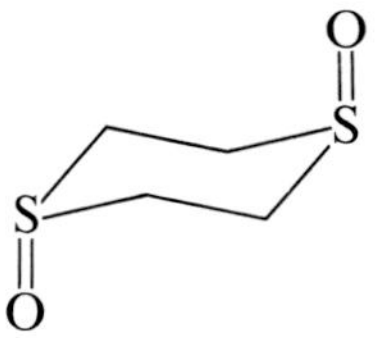   (23) 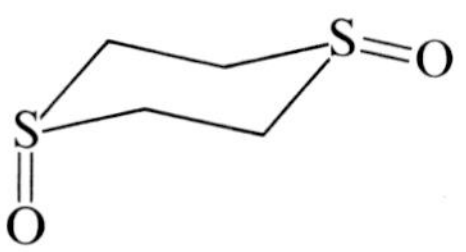   (24)

R-Value measurements on 1,4-dithian and 1,4-diselenan give values of 3.38 and 3.49 respectively.[44] These R values are considerably greater than expected for a normal chair where R is ca 2.0, and are indicative of a puckered chair conformation. This result is in agreement with the X-ray crystallographic work mentioned above,[35-43] which shows that in the absence of specific interactions the 1,4-dithian ring is considerably more puckered than cyclohexane.

1,2,3-Trithian

Conformational equilibria,[45] and ring inversion barriers[46] in some derivatives of 1,2,3-trithian (25) have been reported based on ^{1}H NMR studies. In the 5-methyl-5-alkyl series (26) when R = ethyl, n-propyl, isopropyl or s-butyl the marginally preferred conformations seem to have R axial, whilst when R = isobutyl, neopentyl or phenyl the preferred conformations have R equatorial.

(25)   Me R

                  (26)

1,3,5-Trithian

As early as 1937 X-ray diffraction showed 1,3,5-trithian (27) to have a chair conformation.[47] The ring inversion barrier has been measured to be 11.1 kcal mol^{-1}.[48] In the 2-position of the ring the polar substituents MeS, EtS, isoPrS, and ArS have an axial tendency due to the anomeric effect but t-BuS prefers to be equatorial.[49] Similarly 2-CO_2H prefers to be axial.[30]

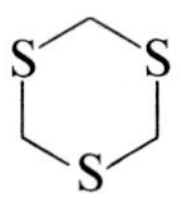

(27)

1,2,4,5-Tetrathian

The most striking feature of the conformational analysis of this system is that in many cases the twist form is favoured over the chair conformation.[50-54]

For the tetramethyl derivative (28) in solution at $-15°$ the chair:twist ratio is $1.0:2.6$ ($\Delta G = 0.49$ kcal mol^{-1}).[50] Most interestingly, ^{1}H NMR spectra were used to demonstrate the existence of the twist conformation in the solid phase. The barrier to the chair → twist process was found to be ca 16 kcal mol^{-1} giving an estimate for the halflife of either chair or twist in solution at $-80°$ of ca 75 h. Dissolving the crystalline material in carbon disulphide at $-80°$ gave a solution whose ^{1}H NMR spectrum was only consistent with a conformationally homogeneous population of twist conformations. Allowing the solution to warm slowly, converted the spectrum to that of the equilibrium mixture. For the spiro derivative (29, $n = 4$) the chair conformation is the more stable in solution; however for (29, $n = 5$) the twist is again the more stable species.

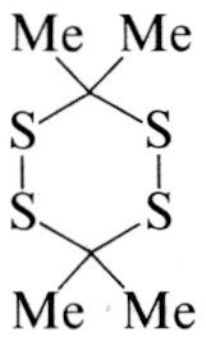

(28)

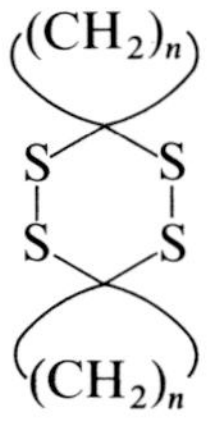

(29)

Bearing in mind the drop in the chair–twist free energy difference on going from cyclohexane to 1,3-dithian, and that the form of the potential barrier about heteroatom–heteroatom bonds is vastly different from those about other "normal" bonds, the twist preference in the tetrathians may be seen as a combination of these effects.

1,2,3,4,5-Pentathian

This compound (30) has been prepared and shows an AB quartet for the ^{1}H spectra of its methylene group, indicating slow ring inversion at ambient temperature ($\Delta G^{\ddagger} >$ ca 15 kcal mol^{-1}).[55] This is in keeping with our ideas of high rotation barriers about heteroatom–heteroatom bonds.

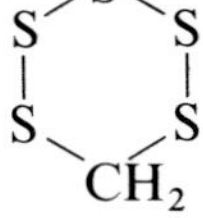

(30)

1,2-Oxathian

Nuclear magnetic resonance studies on 1,2-oxathian-2-oxide (31) and several alkylated derivatives show that there is a considerable, though unquantified, preference for the sulphoxide to be axial even in the presence of a bulky *syn*-axial group such as phenyl.[17, 56]

(31)

1,3-Oxathian

1,3-Oxathian is conformationally a most interesting ring system. It is "schizophrenic" in that at some times it has similar behaviour to 1,3-dithian and at other times to 1,3-dioxan. The 1,3-oxathian system has been most actively investigated by the Finnish group led by Pihlaja, although others have made significant contributions.

Hydrogen-1 NMR spectra show that the ring has the expected chair conformation.[19, 57–61] Coupling constant measurements suggest that the oxygen side of the ring is slightly more puckered than in 1,3-dioxan, and the sulphur side slightly flatter than in 1,3-dithian.[60]

Two claims have been made that there are two chair conformations of the 1,3-oxathian ring that may be constructed with the aid of models: ". . . one with the largest puckering on the *O*-side, the other one being more puckered on the *S*-side."[19, 59] The current author and those other workers in the field whom he has contacted, have been unable to reproduce these findings. What is, however, certain is that due to the dissymmetry of the ring, two enantiomeric chair conformations exist that may be interconverted by a ring inversion process (Fig. 6.3). It is conceivable that this, combined with distorted models, was the cause of the previous confusion.

Careful studies of epimerization equilibria in 1,3-oxathians as models for conformational equilibria were carried out by Pasanen and Pihlaja.[62, 63] The results are presented in Table 6.2. From these results several interesting and constructive comparisons with the results for 1,3-dioxan and 1,3-dithian (Table 6.1) may be made. The conformational energy differences of alkyl substituents at position 2 do not differ appreciably from each other and are similar to, but slightly greater than the mean values for the same interactions in the dioxan and dithian rings. The energy difference for a

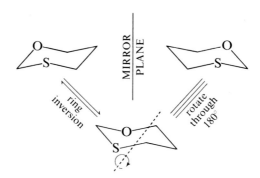

Fig. 6.3   Removal of the dissymmetry of 1,3-oxathian by ring inversion.

4-methyl group is similar to, but marginally larger than that found in 1,3-dithian. In this case the molecule behaves in a 1,3-dithian-like fashion. The energy difference of a 6-methyl group is very close to the value found in 1,3-dioxans. Here the molecule behaves in a 1,3-dioxan-like manner. The energy difference of a 5-methyl group is smaller than those in the dioxan and dithian rings. This may arise from a distortion of the oxathian ring.

(32a)   (32b)

The preferred conformations observed for some 1,3-oxathians with *trans*-dimethyl groups in a 1,3-relationship are in agreement with the results in Table 6.2. Thus the *trans*-2,4-dimethyl derivative exists with 92 % 4-axial (32a) and 8 % 2-axial (32b). The proportion of the 2-axial form increases to 37 % in the *trans*-2,6-dimethyl derivative (33a) with 63 % 6-axial (33b). Finally the 4-axial form (34a) predominates to 88 % in the *trans*-4,6-dimethyl derivative.[57,60,62]

TABLE 6.2   Conformational free energy differences of alkyl substituents in the 1,3-oxathian ring (kcal mol^{-1}) (from Refs. 62–65).

2 Me	3.25	2 Et	3.25	2 *i*-Pr	3.55
4 Me	1.77	—		—	
5 Me	0.68–0.74	—		—	
6 Me	2.94	—		—	

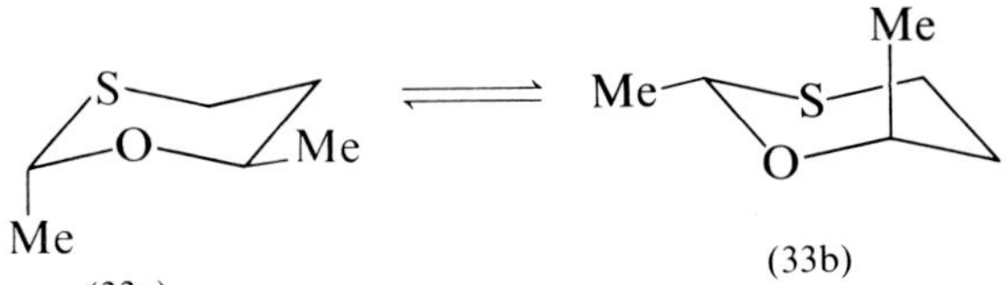

(33a) (33b)

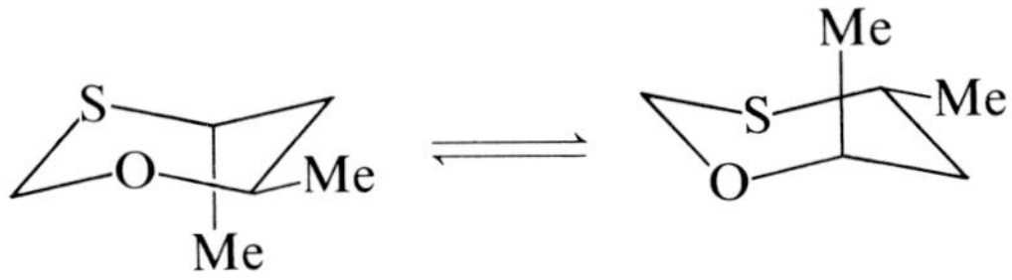

(34a) (34b)

A similar strong preference for the 4-axial conformation is found in trans-4,6-dimethyl-1,3-oxaselenan (35).[66]

Ring inversion processes in 1,3-oxathian derivatives have been investigated and free energies of activation span the range 8.0–11.6 kcal mol^{-1}.[67]

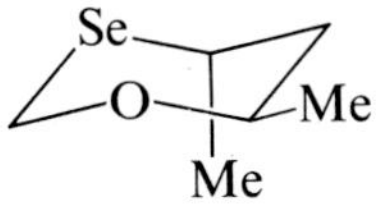

(35)

The chair–twist energy difference in 1,3-oxathian has been estimated as a result of chemical equilibration experiments on 2-t-butyl-2,6-dimethyl-1,3-oxathian (36).[68] The isomer with the methyl groups cis cannot exist in a chair conformation as it would either have an axial 2-t-butyl group or an intolerable 2,6-syn-dimethyl interaction (36b). Consequently it adopts the twist conformation (36c). Correcting the observed enthalpy and entropy differences between the diastereoisomers gives an estimate of 5.5 ± 0.5 kcal mol^{-1} for the free energy difference.[68] Again this value is almost half way between the values for 1,3-dithian and 1,3-dioxan.

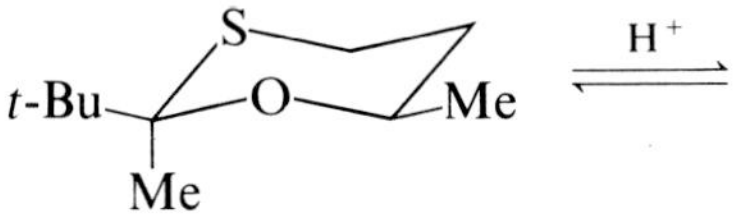

(36a)

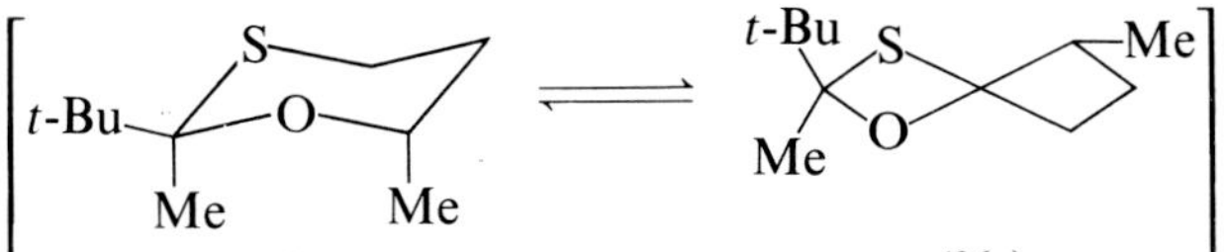

(36b) (36c)

1,4-Oxathian

What little is known about the conformational analysis of 1,4-oxathian derivatives reveals more information about *gauche*–repulsive interactions between sulphur and electronegative substituents on the ring.

X-Ray crystallography shows *trans*-2,3-dichloro-1,4-oxathian to have a chair conformation which is flattened on the chlorine-bearing side with the chlorines leaning out (Fig. 6.4).[69]

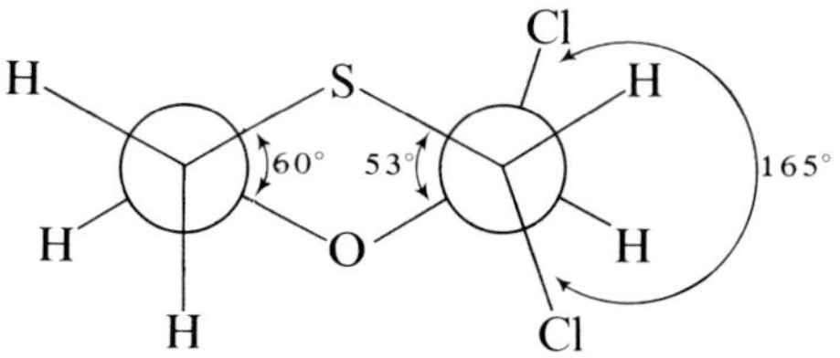

Fig. 6.4   Partial molecular geometry in *trans*-2,3-dichloro-1,4-oxathian.

Nuclear magnetic resonance coupling constant measurements on several derivatives show that in solution chlorinated 1,4-oxathians prefer to have axial halogen atoms.[70]

A quantitative study of conformational equilibria in 1,4-oxathians with electronegative substituents undertaken in Russia[71] and Canada[72–74] shows that these substituents go axial much more readily next to sulphur, where their *gauche* interaction is with oxygen, than when they are next to oxygen and suffer a *gauche* interaction with sulphur. Thus, for 2-methoxy-1,4-oxathian, 50% of the molecules have an axial *O*-methyl group (37) whereas the 3-acetoxy derivative has 90% of its molecules with an axial *O*-acetate

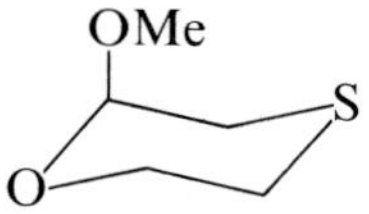

(37)

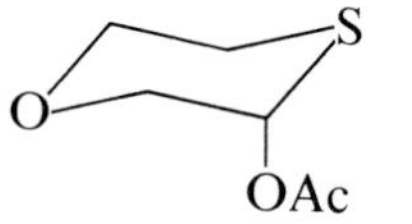

(38)

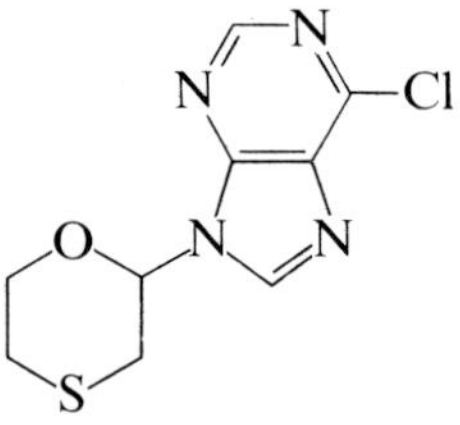

(39)

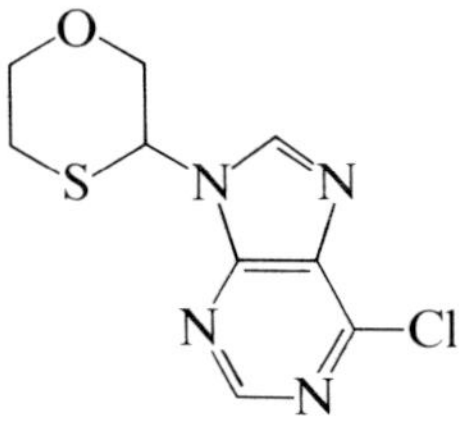

(40)

group (38).[72] Analogous results are observed in the nucleoside analogues (39) and (40).[72-74] In the 2-substituted derivative (39), ca 25% of the molecules have an axial purine group. In the 3-substituted derivative (40) the

(41)

proportion is ca 95%. Clearly the anomeric and *gauche*–repulsive effects are operating in these compounds.

The axial conformation of the 4-oxide (41) is 0.68 kcal mol^{-1} more stable than the equatorial.[75]

1,3,2-Dioxathian

The most studied compounds in this series are derivatives of trimethylene sulphite (42). The weak axial preference for axial O in sulphoxides,[12,13] which is found to be enhanced in the 1,2-oxathian-2-oxides,[17,56] is also seen very strongly in the cyclic sulphites.[76-81]

(42)   (43)

The ring inversion barrier of the parent compound in this series has been measured and found to be 11.2 kcal mol^{-1}.[82] This barrier and those in related compounds are invariably higher than in the related 1,3-dioxans reflecting the larger torsional potential about heteroatom–heteroatom bonds.

Rings Containing Sulphur and Nitrogen

In tetrahydro-1,3-thiazine (43) as in tetrahydro-1,3-oxazine, the N—H has been shown to be axial.[83] This preference extends into the N-alkyl derivatives where the N-methyl derivative has been shown to be 0.7 kcal mol^{-1} more stable when axial.[84] Similar axial preferences have been shown in the N-alkyldihydro-1,3,5-dithiazines (44) and in the N,N'-dialkyltetrahydro-1,3,5-thiadiazines (45).[85] However, this preference is not sufficient to convert the preferred *trans* conformation of 1-aza-3-thiabicyclo[4,4,0]-decane (46) to the *cis*-fused form.[86] This preference for N-alkyl groups in

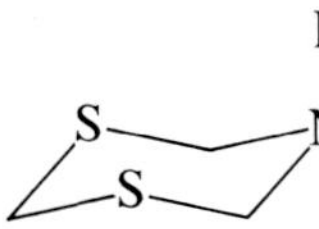

(44) 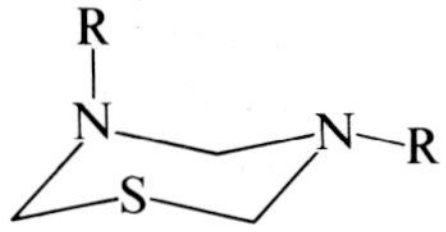   (45)

(46) 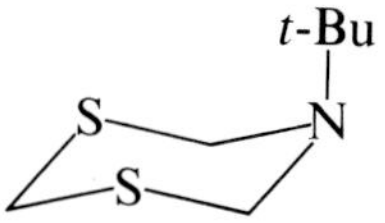   (47)

1,3-relationship to sulphur to go axial, almost certainly arises from an anomeric effect coupled with the absence or loss of hindering *syn*-axial hydrogens. This is shown even more dramatically by the axial preference of the *t*-butyl group in the dithiazine (47) ($\Delta G = 0.8$ kcal mol^{-1}).[87] Reports of the conformations of 1,4-thiazines[88] and 1,4,2,3-dithiadiazines[89] have appeared.

Phosphorus-Containing Rings

Introduction

Phosphorus-containing rings have proved a popular vehicle for conformational studies. The reasons for this include the biological importance of some derivatives, especially phosphate esters, and also more importantly their ease of synthesis and the many varied methods available for their study. Nuclear magnetic resonance methods have proved the most useful and popular because three nuclear species may be studied in these rings: ^{1}H, ^{13}C, and ^{31}P. X-Ray diffraction, dipole moments and infrared spectra have also played important roles.

Most of the work in this area developed during the early 1970s. A review of the stereochemistry of phosphorus compounds published in 1969 covers the information then available on the conformations of phosphorus compounds in a few brief pages.[90a] Since then, several research groups have made substantial contributions including those led by W. G. Bentrude. L. D. Quin, J. B. Robert, and J. G. Verkade[90]

The chemistry of phosphorus compounds is complicated by the variable valency of phosphorus and the consequent variety of the number and type of ligand present on phosphorus. Hence the conformational analysis of phosphorus-containing rings is somewhat more complex, in its variations, than for the rings we have previously considered.

Phosphorus-containing rings show several similar conformational trends to those of sulphur, including axial preferences on phosphorus of a wide variety of substituents, so we consider them in the same chapter as sulphur.

Phosphorinan

Several X-ray crystallographic studies on derivatives of this ring system have been recorded.[91-94] In all cases the rings are in chair conformations which appear to be slighly flattened at the phosphorus end. In the 1-phenyl-4,4-dimethoxy derivative (48) X-ray diffraction shows a chair conformation with an axial phenyl group.[91] The 4-*gem*-Dimethoxy group shows the expected *gauche–gauche* arrangement to minimize its internal anomeric interactions (Chapter 4) and the ring is flattened at the phosphorus atom. For the parent ketone (49) of this ketal the *P*-phenyl group is again found to be axial.[94] The 2-thio derivative (50) co-crystallizes in two conformations (50a and 50b).[93]

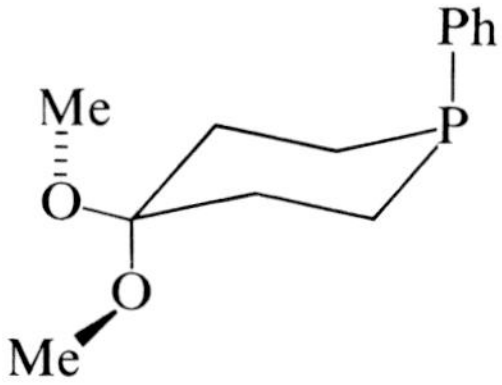

(48)

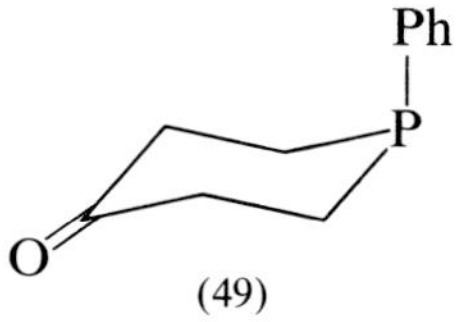

(49)

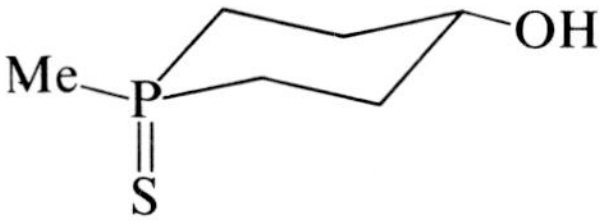

(50a)

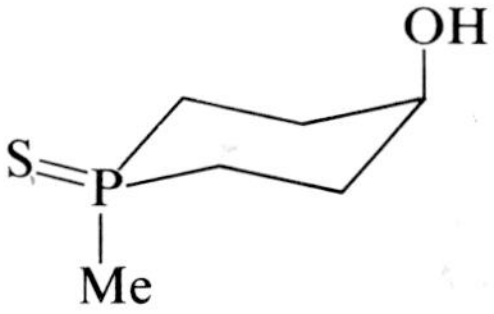

(50b)

This is probably the only known example of two different chair conformations of a single six-membered ring compound co-crystallizing. In this and related compounds the flattening of the ring around phosphorus is seen in the ring dihedral angles of 45–52° around P, whereas the other ring dihedral angles are 55–59° (C(2)–C(3)) and 60–64° (C(3)–C(4)).[93]

Analysis of the ^{1}H NMR spectrum of phosphorinan (51) shows vicinal couplings to the hydrogen on phosphorus of 12.0 and 2.5 Hz.[95] These are

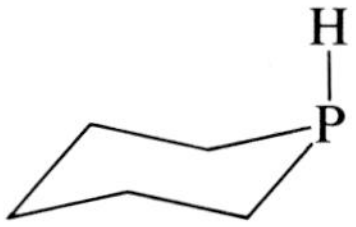

(51)

clearly consistent with an axial P—H bond. A similar conclusion was reached on similar grounds for the 1-thio-derivative (52).

Low-temperature ^{31}P NMR spectra on 1-methylphosphorinan[96] and on other 1-alkyl and 1-phenyl phosphorinans have been reported by Featherman and Quin.[97] At 27° the phosphorus resonances are sharp singlets which split into unequal doublets on lowering the temperature due to slowing of ring inversion. Identification and integration of the peaks shows that the equatorial form (53a) predominates at low temperatures (ca −100°) and is lower in enthalpy by 0.6 to 0.7 kcal mol^{-1}, but that there is a substantial entropy difference of 3–4 cal K^{-1} favouring the axial conformation (53b), which consequently becomes the predominant form at ambient temperatures. The ring flattening at phosphorus and the long C—P bonds are responsible for lowering the energy difference relative to methyl cyclohexane.

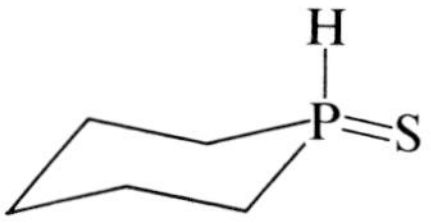

(52)

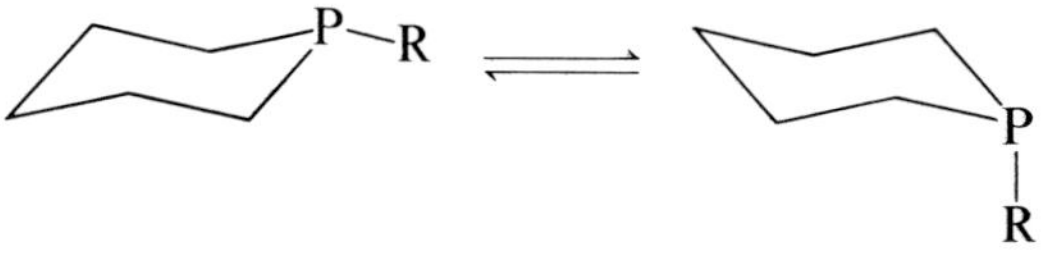

(53a)   (53b)

R = Me, Et, i-Pr, Ph

The axial preference of 2-methyl, ethyl, and phenyl substituents at ambient temperatures was subsequently confirmed by ^{13}C NMR studies.[98] This work, however, shows 2-isopropyl and 2-t-butyl groups to be predominantly equatorial.

Reports of the conformations of some 4-phosphorinanol derivatives (54),[99, 100] show that cis and trans isomers generally differ in the configuration of the phosphorus atom (e.g. 54a and 54b) rather than in the configuration at the hydroxylic carbon atom. This is in accordance with the conformational free energy difference on phosphorus being smaller than that of the hydroxyl group.

In some chemical equilibrations on 4-phospha-2,6-dimethylcyclo-hexanones (55) as models for conformational equilibria on the ring carbon atoms, it was found that the position of the equilibrium varied little with the

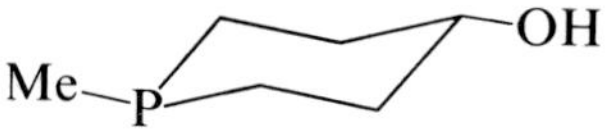

(54a) 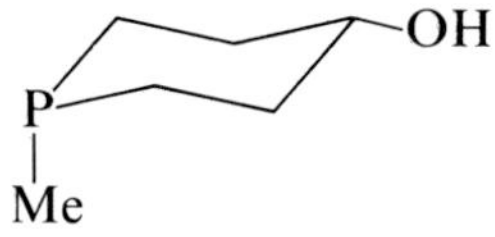   (54b)

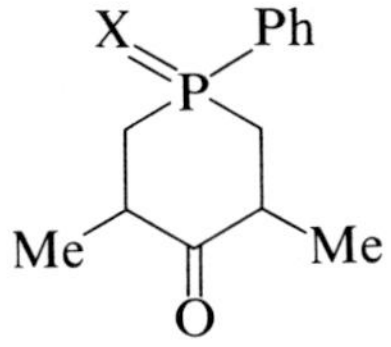

(55)

substitution pattern of the phosphorus.[101] The diequatorial conformation
was favoured by 1.08 to 1.14 kcal mol^{-1}.

1,4-Diphosphorinan

Tetraphenyl diphosphoranium dibromide (56) has been shown to have a
chair conformation by X-ray diffraction.[102]

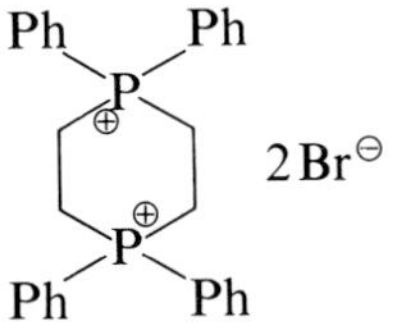

(56)

1,2-Oxaphosphorinan

The conformations of several compounds in this series (57, 58, 59) have
been studied by Bergesen's group in Norway.[103-107] In these reports there
is some confusion over nomenclature. In some cases the nomenclature *cis*
refers to the relationship between alkyl group on carbon and the oxo group
on phosphorus. In others it refers to the relationship between the alkyl
group and the ethoxy group on phosphorus. Moreover the means of identification of the isomers, which is based on very small differences in P=O
stretching frequencies (ca 5 cm^{-1}) and on ^{1}H chemical shift measurements,
is not above question. Nevertheless, it appears from these reports that a

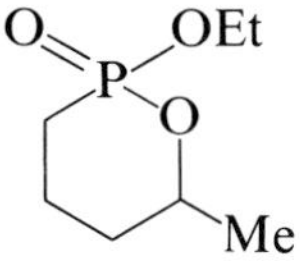

(57) 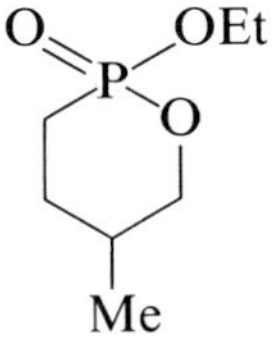   (58) 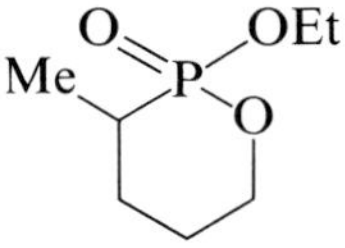   (59)

2-alkoxy group has a substantial axial preference. Thus in, for example, the 3-methyl derivative the equilibrium between the isomers can be established by TFA at elevated temperatures. It seems that the preferred conformations of each isomer are (57a) and (57b), and that the isomer (57a) does not exist to any appreciable extent in the ring inverted form (57c).

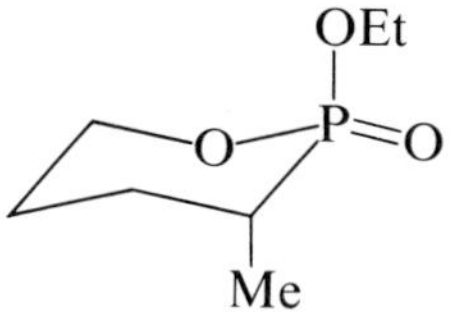

(57a) 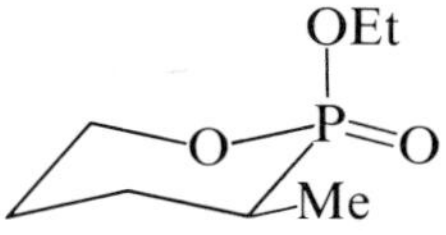   (57b) 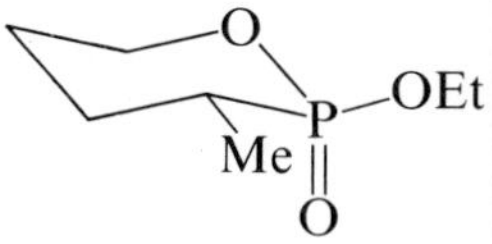   (57c)

1,3,2-Dioxaphosphorinan

General

This is by far the most extensively investigated six-membered ring system containing phosphorus. Many reports of X-ray crystallographic studies of derivatives in this series have been made.[108-115] Compounds studied, which are drawn in the formulae in their observed conformations, include: (60),[108] (61),[109] (62),[110] (63),[111] (64),[112] (65),[113] (66),[114] and (67),[115] Assuming, as is

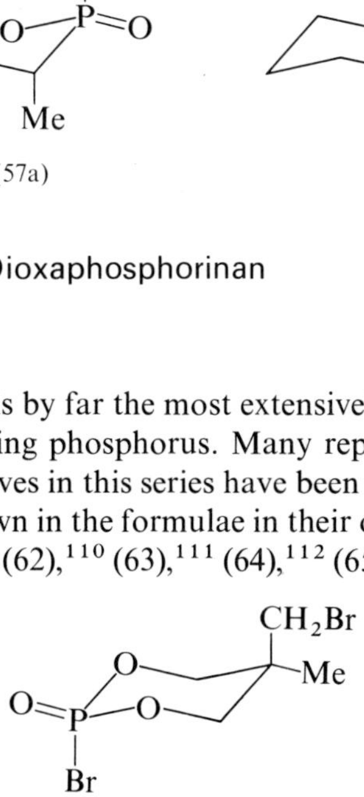

(60) 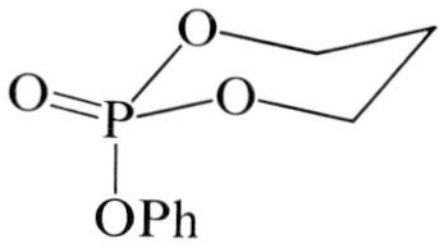   (61)

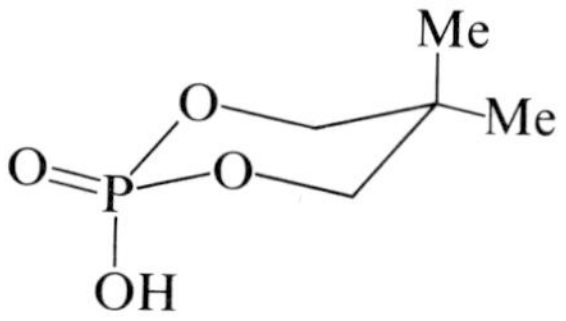

(62) 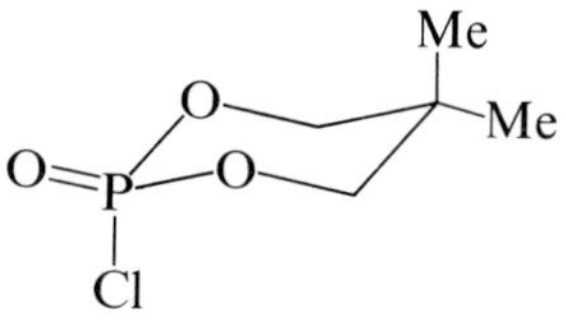   (63)

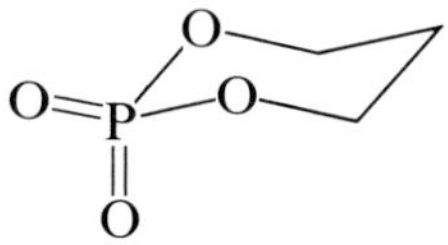

(64) 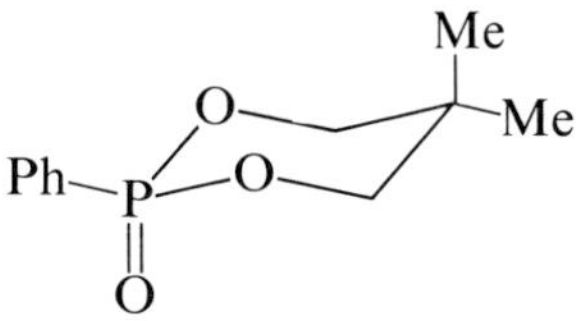   (65)

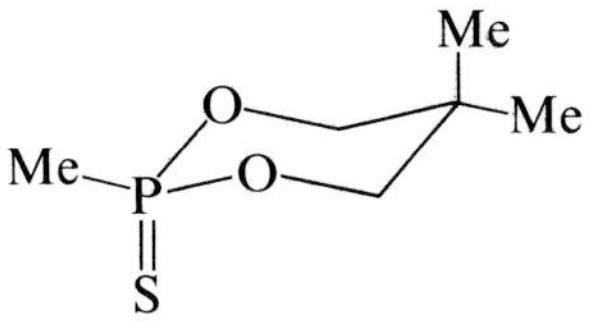

(66) 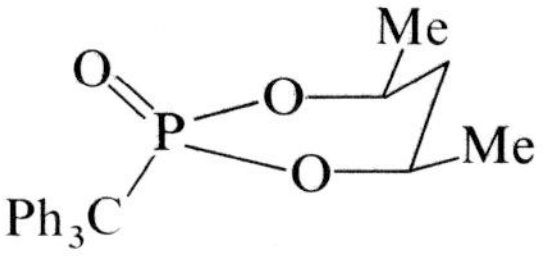   (67)

often the case, that the preferred conformation in the crystal is the same as the preferred conformation in solution, these structures suggest several conformational preferences that are confirmed by other methods. Firstly the electronegative substituents, —OR, Cl, =O, =S, like to be axial on phosphorus. Secondly, the axial preference of —OR is greater than that of =O. In most cases the rings are observed to be somewhat distorted from "true" chairs, being flatter at the end containing phosphorus. This is particularly true in (67),[115] where the ring is better described as a half-chair (or *chaise longue*!) and is much more distorted than in the related structures.

Three ligands on phosphorus

The evidence here, from a variety of methods, shows that most groups on phosphorus prefer to be axial (68).[116–129] Free energy differences on phosphorus measured at $25°$ are: Cl, 1.5; OMe, 1.4; Ph, 1.3; Me, 0.99; *i*-Pr, 0.63; *t*-Bu, -1.5; Me_2N, -0.94; MeHN, -0.12 kcal mol^{-1}.[116] Apart from the case of the amino substituents, electronegative groups and the smaller alkyl groups prefer to be axial, reminiscent of observations in sulphur-containing rings, and the phosphorus-containing rings discussed above.

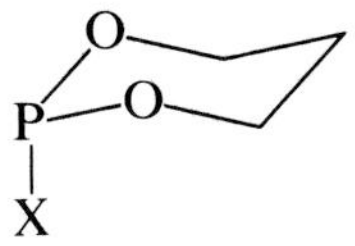

(68a)   (68b)

X = Cl, OMe, CH_3, *i*-Pr, Ph   X = *t*-Bu, NMe_2, NHMe

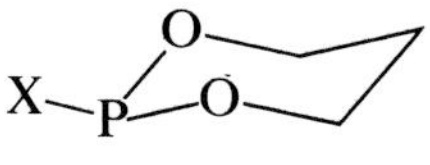

Several different influences are probably at work in determining these axial preferences. For thian and phosphorinan it is suggested that the axial S or P substituent is located at a distance from the *syn*-axial hydrogens that causes weak attractions in the minimum of the van der Waals' potential energy curve. This will be brought about by the length of the C—P and C—S bonds. When oxygen atoms surround P or S in the ring the anomeric effect can come into play for electronegative substituents. This is seen to some extent even in 2-methoxy-1,3-dioxan, although here the axial preference is reduced by severe non-bonded interactions (Chapter 4). Finally, rotational preferences about the O—P and O—S bonds are probably very different from those about bonds involving carbon. Our knowledge of, for example, the conformational preferences about the closely related O—N bond shows that the conformation with an axial substituent should have less *torsional* strain about the O—X bond than the conformation with the substituent equatorial. The interplay between these effects in the 1,3,2-dioxaphosphorinans has been discussed in detail by Bentrude.[116]

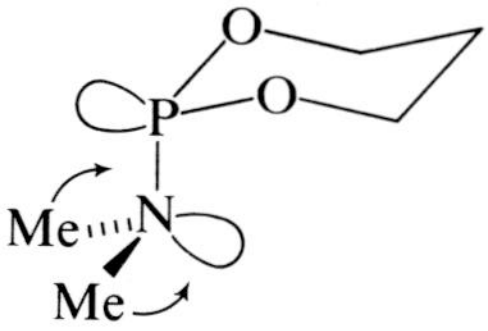

(69a) 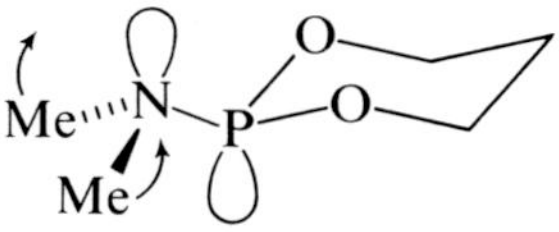   (70)

The equatorial preference of a *t*-butyl group is clearly steric in origin. The equatorial preference of amino substituents has been suggested to arise from steric inhibition of dπ–pπ bonding. For an axial dialkylamino substituent dπ–pπ bonding is hindered (69) to a much greater extent than when the substituent is equatorial (70). In accordance with this view the equatorial preference is greater with the bulkier NMe_2 group, and smaller with the less sterically demanding NHMe group.[126]

Four ligands on phosphorus

A similar set of conformational tendencies to those described above are observed when the oxidation state of phosphorus is increased and four ligands are present on phosphorus. For 2-oxo-2-alkoxy derivatives the axial alkoxy group is favoured (71).[130–134] For 2-oxo-2-alkyl derivatives an axial oxo group is found (72).[133–138] In derivatives with 2-oxo-2-amino substitution the oxo group goes axial (73).[133,134,139] An axial 2-diphenyl-phosphino group is preferred to an axial 2-oxo group (74).[140] Similar tendencies are found in 2-thio-2-substituted derivatives (75).[141] There is some evidence that non-chair conformations are important in this series.

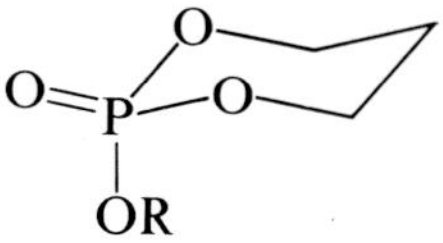

(71) 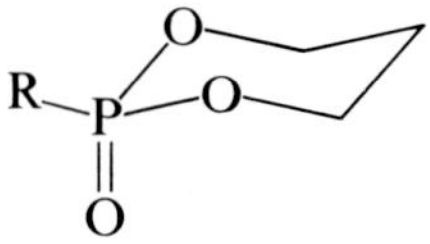   (72) 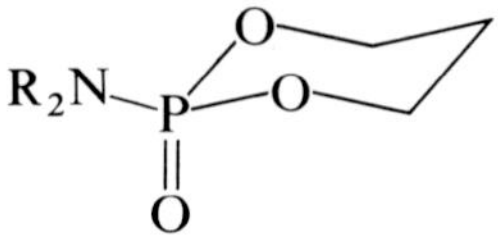   (73)

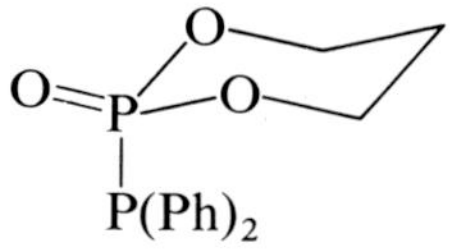

(74) 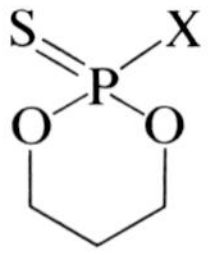   (75)

1,3,2-Diazaphosphorinan

A small amount of work on compounds in this series has been reported. Hydrogen-1 and ^{31}P NMR spectra show the rings to be in chair conformations with their N-substituents probably equatorial.[142–143] The endocyclic $PNCH$ couplings vary greatly with the nature of the P-substituent, suggesting that hybridization on nitrogen, and consequently the ring geometry, is very dependent upon the electronegativity of the P-substituent.[142]

1,3,2-Dithiaphosphorinan

The axial preference of groups on phosphorus noted in the above compounds is also seen in this series. For the 5,5-dimethyl derivatives (76) the axial position is preferred when R = Me, Et, Ph, and OMe but not when R = t-Bu.[144] Nuclear magnetic resonance studies on the 2-phenyl derivative (77) show that coupling constants in the trimethylene chain are very similar to those in 2-phenyl-1,3-dithian (78) indicating a very similar conformation for the S—$(CH_2)_3$—S unit in both ring systems.[145,146]

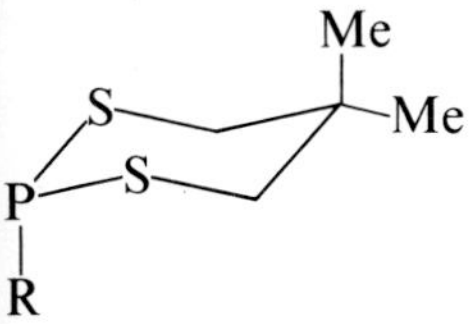

(76) 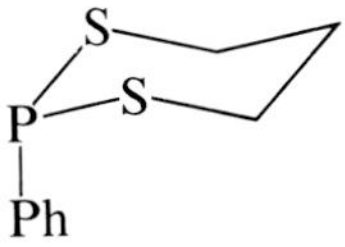   (77) 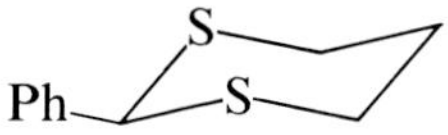   (78)

1,3,2-Oxazaphosphorinan

This ring system is of importance in that a derivative is widely used as an anticancer drug under the commercial name of cyclophosphoramide (79). Because of this, several X-ray crystallographic studies have appeared of what might otherwise have been a little-studied system.[147–155] Reports of conformational equilibria in derivatives of this ring show very similar tendencies to the 1,3,2-dioxaphosphorinans.[156–158]

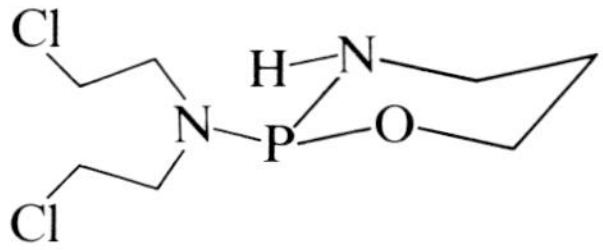

(79)

References

1.   N. S. Zefirov and I. V. Kazimirchik. (1974). *Russian. Chem. Rev.* **43**, 107.
2.   J. B. Lambert, R. G. Keske and D. K. Weary. (1967). *J. Am. Chem. Soc.* **89**, 5921.
3.   J. B. Lambert, C. E. Mixan and D. H. Johnson. (1973). *J. Am. Chem. Soc.* **95**, 4634.
4.   N. L. Allinger and M. J. Hickey. (1975). *J. Am. Chem. Soc.* **97**, 5167.
5.   R. L. Willer and E. L. Eliel. (1977). *J. Am. Chem. Soc.* **99**, 1925.
6.   R. Gerdil. (1974). *Helv. Chim. Acta.* **57**, 489.
7a. E. L. Eliel, R. L. Willer, A. T. McPhail and K. D. Onan. (1974). *J. Am. Chem. Soc.* **96**, 3021.
7b. M. J. Cook, H. Dorn and A. R. Katritzky. (1968). *J. Chem. Soc.* (*B*) 1467.
8.   G. Barbarella, P. Dembech, A. Garbesi and A. Fava. (1976). *Tetrahedron.* **32**, 1045; (1976). *Org. Mag. Res.* **8**, 469, 108.
9.   R. L. Willer and E. L. Eliel. (1977). *Org. Mag. Res.* **9**, 285.
10. O. Hofer and E. L. Eliel. (1973). *J. Am. Chem. Soc.* **95**, 8045.
11. See for example, F. G. Riddell. (1967). *Quart. Rev.* **21**, 364.
12. J. B. Lambert and R. G. Keske. (1966). *J. Org. Chem.* **31**, 3429.
13. G. W. Buchanan and T. Durst. (1975). *Tetrahedron Lett.* 1683.
14. J. B. Lambert, C. E. Mixan and D. H. Johnson. (1972). *Tetrahedron Lett.* 4335.
15. J. B. Lambert, D. S. Bailey and C. E. Mixan. (1972). *J. Org. Chem.* **37**, 377.
16. O. Foss, K. Johnsen and T. Reistad. (1964). *Acta Chem. Scand.* **18**, 2345.
17. D. N. Harpp and J. G. Gleason. (1971). *J. Org. Chem.* **36**, 1314.
18. H. T. Kalff and C. Romers. (1966). *Acta Cryst.* **20**, 490.
19. J. Gelan, G. Swaelens and M. Anteunis. (1970). *Bull. Soc. Chim. Belges.* **79**, 321.
20. F. A. Carey, P. M. Smith, R. J. Maher and R. F. Bryan. (1977). *J. Org. Chem.* **42**, 961.
21. A. T. McPhail, K. D. Onan and J. Koskimies. (1976). *J.C.S. Perkin II* 1004.
22. E. L. Eliel and R. O. Hutchins. (1969). *J. Am. Chem. Soc.* **91**, 2703.
23. K. Pihlaja. (1974). *J.C.S. Perkin II* 890.
24. K. Pihlaja and H. Nikander. (1977). *Acta Chem. Scand.* **B31**, 265.
25. E. L. Eliel, V. S. Rao and F. G. Riddell. (1976). *J. Am. Chem. Soc.* **98**, 3583.
26. K. Pihlaja and B. Bjoerkquist. (1977). *Org. Mag. Res.* **9**, 533.
27. F. A. Carey, O. D. Dailey and W. C. Hutton. (1978). *J. Org. Chem.* **43**, 96.
28. R. J. Abraham and W. A. Thomas. (1965). *J. Chem. Soc.* 335.
29. A. Luttringhaus, S. Kabuss, H. Prinzbach and F. Langenbucher. (1962). *Annalen* **653**, 195.
30. K. Arai, H. Iwamura and M. Oki. (1975). *Bull. Chem. Soc. Jpn* **48**, 3319.
31. S. A. Khan, J. B. Lambert, O. Hernandez and F. A. Carey. (1975). *J. Am. Chem. Soc.* **97**, 1468.
32. M. J. Cook and A. P. Tonge. *Tetrahedron Lett.* (1973). 849; *J.C.S. Perkin II* (1974). 767.
33. L. van Acken and M. Anteunis. (1974). *Tetrahedron Lett.* 225.
34. K. Arai and M. Oki. (1975). *Tetrahedron Lett.* 2183.
35. R. E. Marsh. (1955). *Acta Cryst.* **8**, 91.
36. S. Y. Chao and J. D. McCullough. (1960). *Acta Cryst.* **13**, 727.
37. T. Bjorvatten and O. Hassel. (1961). *Acta Chem. Scand.* **15**, 1429.
38. T. Bjorvatten. (1966). *Acta Chem. Scand.* **20**, 1863.
39. O. Holmesland and C. Romming. (1966). *Acta Chem. Scand.* **20**, 2601.

40. H. M. M. Shearer. (1959). *J. Chem. Soc.* 1394.
41. H. Montgomery. (1960). *Acta Cryst.* **13**, 381.
42. H. T. Kalff and C. Romers. (1965). *Acta Cryst.* **18**, 164.
43. H. T. Kalff and C. Romers. (1966). *Rev. Trav. Chim.* **85**, 198.
44. J. B. Lambert. (1967). *J. Am. Chem. Soc.* **89**, 1836.
45. G. Goor and M. Anteunis. (1975). *Bull. Soc. Chim. Belges* **84**, 337.
46. G. Goor and M. Anteunis. (1975). *Heterocycles* **3**, 363.
47. N. F. Moerman and E. H. Wibenga. (1937). *Z. Krist.* **97**, 323.
48. J. E. Anderson. (1971). *J. Chem. Soc.* (*B*) 2030.
49. M. Oki, T. Endo and T. Sugawara. (1975). *Bull. Chem. Soc. Jpn* **48**, 2496.
50. C. M. Bushweller. *J. Am. Chem. Soc.* (1967). **89**, 5978; (1968). **90**, 2450; (1969). **91**, 6019.
51. C. H. Bushweller, J. Golini, G. V. Rao and J. W. O'Neill. (1970). *J.C.S. Chem. Comm.* 51; (1970). *J. Am. Chem. Soc.* **92**, 3055.
52. C. H. Bushweller, G. U. Rao and F. H. Bissett. (1971). *J. Am. Chem. Soc.* **93**, 3058; (1971). *Int. J. Sulfur Chem.* 216.
53. A. Fredga. (1958). *Acta Chem. Scand.* **12**, 891.
54. C. H. Bushweller, G. Bhat, L. J. Letendre, J. A. Brunelle, H. S. Bilofsky, H. Ruben, D. H. Templeton and A. Zalkin. (1975). *J. Am. Chem. Soc.* **97**, 65.
55. F. Feher, B. Degen and B. Sohnen. (1968). *Angew. Chem.* (Int. ed English), **7**, 301.
56. G. W. Buchanan, N. K. Sharma, F. De-Reinach-Hirtzbach and T. Durst. (1977). *Can. J. Chem.* **55**, 44.
57. J. Gelan and M. Anteunis. (1968). *Bull. Soc. Chim. Belges* **77**, 423.
58. J. Gelan and M. Anteunis. (1968). *Bull. Soc. Chim. Belges* **77**, 447.
59. N. de Wolf and H. R. Buys. (1970). *Tetrahedron Lett.* 551.
60. P. Pasanen. (1972). *Suomen Kem.* **B45**, 363.
61. Y. Allingham, T. A. Crabb and R. F. Newton. (1971). *Org. Mag. Res.* **3**, 37.
62. P. Pasanen and K. Pihlaja. (1972). *Tetrahedron* **28**, 2617.
63. P. Pasanen. (1974). *Finn. Chem. Lett.* **1**, 49.
64. P. Pasanen, Doctoral Thesis, University of Turku, Finland, 1974.
65. J. Jalonen, P. Pasanen and K. Pihlaja. (1973). *Org. Mass. Spectrom.* **7**, 949.
66. A. Geens, M. Anteunis, F. De Pessemier, J. Fransen and G. Verhege. (1972). *Tetrahedron* **28**, 1097.
67. H. Friebolin, H. G. Schmid, S. Kabuss and W. Faisst. (1969). *Org. Mag. Res.* **1**, 67. See also, J. E. Anderson, (1974). *Fortschritte* **45**, 139.
68. K. Pihlaja and P. Pasanen. (1974). *J. Org. Chem.* **39**, 1948.
69. N. de Wolf, C. Romers and C. Altona. (1967). *Acta Cryst.* **22**, 715.
70. N. de Wolf, P. W. Henniger and E. Havinga. (1967). *Rec. Trav. Chim.* **86**, 1227.
71. N. S. Zefirov, V. S. Blagoveshchensky, I. V. Kazimirchik and N. S. Surova. (1971). *Tetrahedron* **27**, 3111.
72. W. A. Szarek, D. M. Vyas, A.-M. Sepulchre, S. D. Gero and G. Lukacs. (1974). *Canad. J. Chem.* **52**, 2041.
73. W. A. Szarek, D. M. Vyas and B. Achmatowicz. (1975). *Tetrahedron Lett.* 1553.
74. W. A. Szarek, D. M. Vyas and B. Achmatowicz. (1975). *J. Heterocyclic Chem.* **12**, 123.
75. D. M. Frieze and S. A. Evans. (1975). *J. Org. Chem.* **40**, 2690.
76. C. Altona, H. J. Geise and N. C. Romers. (1966). *Rec, Trav. Chim.* **85**, 1197.
77. H. F. van Woerden and E. Havinga. (1967). *Rec. Trav. Chim.* **86**, 341, 343.
78. F. J. Mustoe and J. L. Hencher. (1972). *Canad. J. Chem.* **50**, 3892.

79. P. Maroni, L. Cazaux, G. Chassaing, I. Prefzner and T. le Trang. (1975). *Bull. Soc. Chim. France* 1258.
80. P. Maroni, L. Cazaux, J.-P. Gorrichon, P. Tisnes and J. G. Wolf. (1975). *Bull. Soc. Chim. France* 1253.
81. C. H. Green and D. G. Hellier. (1972). *J.C.S. Perkin II* 458; (1973). 243.
82. G. Wood, R. M. Srivatsava and B. Adlam. *Canad. J. Chem.* (1973). **51**, 1200; G. Wood and R. M. Srivatsava. *Tetrahedron Lett.* (1971). 2937.
83. M. J. Cook, R. A. Y. Jones, A. R. Katritzky, M. Moreno-Mañas, A. C. Richards, A. J. Sparrow and D. L. Trepanier. (1973). *J.C.S. Perkin II* 325.
84. V. J. Baker, A. R. Katritzky, F. M. S. Brito-Palma and L. A. Angiolini, unpublished results, personal communication.
85. L. Anglioline, R. P. Duke, R. A. Y. Jones and A. R. Katritzky. (1972). *J.C.S. Perkin II* 674; V. J. Baker, I. J. Ferguson, A. R. Katritzky, R. C. Patel and S. Rahimi-Rastgoo. (1978). *J.C.S. Perkin II* 378.
86. T. A. Crabb and R. F. Newton. (1970). *Tetrahedron* **26**, 3941.
87. L. Angiolini, R. P. Duke, R. A. Y. Jones and A. R. Katritzky. (1971). *Chem. Comm.* 1308.
88. J. F. Carson, L. M. Boggs and R. E. Lundin. (1970). *J. Org. Chem.* **35**, 1594.
89. K. H. Linke, R. Bimczok and H. Lingmann. (1971). *Angew. Chemie* (Int. ed. English) 10.
90a. M. J. Gallacher and I. D. Jenkins. (1968). *In* "Topics in Stereochemistry", Vol. 3, (Eds Allinger and Eliel), 1. Wiley Interscience, New York.
90b. B. E. Maryanoff, R. O. Hutchins and C. A. Maryanoff. (1979). *Topics Stereochem.* **11**, 186.
91. A. T. McPhail, J. J. Breen, J. H. Sommers, J. C. H. Steele and L. D. Quin. (1971). *Chem. Comm.* 1020.
92. A. T. McPhail, P. A. Liehan, S. J. Featherman and L. D. Quin. (1972). *J. Am. Chem. Soc.* **94**, 2126.
93. L. D. Quin, A. T. McPhail, S. O. Lee and K. D. Onan. (1974). *Tetrahedron Lett.* 3473.
94. A. T. McPhail, J. J. Breen and L. D. Quin. (1971). *J. Am. Chem. Soc.* **93**, 2574.
95. J. B. Lambert and W. L. Oliver. (1971). *Tetrahedron* **27**, 4245.
96. S. I. Featherman and L. D. Quin. (1973). *J. Am. Chem. Soc.* **95**, 1699.
97. S. I. Featherman and L. D. Quin. (1975). *J. Am. Chem. Soc.* **97**, 4349.
98. S. I. Featherman, S. O. Lee and L. D. Quin. (1974). *J. Org. Chem.* **39**, 2899.
99. S. I. Featherman and L. D. Quin. (1973). *Tetrahedron Lett.* 1955.
100. H. E. Shook, Jr. and L. D. Quin. (1967). *J. Am. Chem. Soc.* **89**, 1841.
101. I. D. Blackburne, A. R. Katritzky, D. M. Read, R. Bodalski and K. Pietrusiewicz. (1974). *J.C.S. Perkin II* 1155.
102. J. N. Brown and L. M. Trefanas. (1972). *J. Heterocyclic Chem.* **9**, 35.
103. K. Bergesen and A. Berge. (1970). *Acta Chem. Scand.* **24**, 1844.
104. K. Bergesen. (1970). *Acta Chem. Scand.* **24**, 2019.
105. K. Bergesen and T. Vikane. (1971). *Acta Chem. Scand.* **25**, 1147.
106. K. Bergesen and T. Vikane. (1972). *Acta Chem. Scand.* **26**, 1794.
107. K. Bergesen and A. Berge. (1972). *Acta Chem. Scand.* **26**, 2975.
108. T. A. Beinecke. (1966). *Chem. Comm.* 860.
109. H. J. Geisse. (1967). *Rec. Trav. Chim.* **86**, 362.
110. W. Murayama and M. Kainosho. (1969). *Bull. Chem. Soc. Jpn* **42**, 1819.
111. R. C. G. Killean, J. L. Lawrence and I. M. Magennis. (1971). *Acta Cryst* **B27** 189.

112.   L. Silver and R. Rudman. (1972). *Acta Cryst.* **B28**, 574.
113.   M. Ul-Haque, C. N. Caughlan and W. C. Moats. (1970). *J. Org. Chem.* **35**, 1446.
114.   J. P. Dutasta, A. Grand and J. B. Robert. (1974). *Tetrahedron Lett.* 2655.
115.   M. G. B. Drew, J. Rodgers, D. W. White and J. G. Verkade. (1971). *Chem Comm.* 227.
116.   W. G. Bentrude, H.-W. Tan and K. C. Yee. (1975). *J. Am. Chem. Soc.* **97**, 57.
117.   W. G. Bentrude and H.-W. Tan. (1973). *J. Am. Chem. Soc.* **95**, 4666.
118.   W. G. Bentrude and J. M. Hargis. (1970). *J. Am. Chem. Soc.* **92**, 7136.
119.   K. Bergesen and P. Albriktsen. (1972). *Acta Chem. Scand.* **26**, 1680.
120.   D. W. White, R. D. Bertrand, G. K. McEwen and J. G. Verkade. (1970). *J. Am. Chem. Soc.* **92**, 7125.
121.   J. A. Mosbo and J. G. Verkade. (1973). *J. Am. Chem. Soc.* **95**, 4659.
122.   M. Haemers, R. Ottinger, J. Reisse and D. Zimmerman. (1971). *Tetrahedron Lett.* 461.
123.   M. Haemers, R. Ottinger, D. Zimmemran and J. Reisse. (1973). *Tetrahedron Lett.* 2241; *Tetrahedron* (1973). **29**, 3539.
124.   W. G. Bentrude and K. C. Yee. (1970). *Tetrahedron Lett.* 3999.
125.   W. G. Bentrude, K. C. Yee, R. D. Bertrand and D. M. Grant. (1971). *J. Am. Chem. Soc.* **93**, 797.
126.   W. G. Bentrude and H.-W. Tan. (1972). *J. Am. Chem. Soc.* **94**, 8222.
127.   J. A. Mosbo and J. G. Verkade. (1972). *J. Am. Chem. Soc.* **94**, 8224.
128.   J. A. Mosbo and J. G. Verkade. (1973). *J. Am. Chem. Soc.* **95**, 4659.
129.   A. Cogne, A. G. Guimaraes, J. Martin, R. Nardin, J. B. Robert and W. J. Stec. (1974). *Org. Mag. Res.* **6**, 629.
130.   J. P. Majoral and J. Navech. (1971). *Bull. Soc. Chim. France* **95**, 1331, 2609.
131.   D. W. White, G. K. McEwen, R. D. Bertrand and J. G. Verkade. (1971). *J. Chem. Soc. (B)* 1454.
132.   L. D. Hall and R. B. Malcolm. (1972). *Canad. J. Chem.* **50**, 2092, 2102.
133.   J. A. Mosbo and J. G. Verkade. (1977). *J. Org. Chem.* **42**, 1549.
134.   J. A. Mosbo. (1978). *Org. Mag. Res.* **11**, 281.
135.   K. C. Yee and W. G. Bentrude. (1971). *Tetrahedron Lett.* 2775.
136.   W. G. Bentrude and K. C. Yee. (1972). *Chem. Comm.* 169.
137.   A. J. Dale. (1972). *Acta Chem. Scand.* **26**, 2985.
138.   W. G. Bentrude, H.-W. Tan and K. C. Yee. (1972). *J. Am. Chem. Soc.* **94**, 3264.
139.   R. S. Edmundson. (1972). *J.C.S. Perkin I* 1661.
140.   A. Okruszek, W. J. Stec and R. K. Harris. (1977). *Org. Mag. Res.* **9**, 497.
141.   J. P. Dutasta, A. Grand, J. B. Robert and M. Taieb. (1974). *Tetrahedron Lett.* 2659.
142.   R. O. Hutchins, B. E. Maryanoff, J. P. Albrand, A. Cogne, D. Gagnaire and J. B. Robert. (1972). *J. Am. Chem. Soc.* **94**, 9151.
143.   R. Kraemer and J. Navech. (1971). *Bull. Soc. Chim. France* 3580.
144.   R. O. Hutchins and B. E. Maryanoff. (1972). *J. Am. Chem. Soc.* **94**, 3266.
145.   K. Bergesen. (1975). *Acta Chem. Scand.* **A28**, 567.
146.   J. Martin and J. B. Robert. (1975). *Org. Mag. Res.* **7**, 76.
147.   I. L. Karle, J. M. Karle, W. Egan, G. Zon and J. A. Brandt. (1977). *J. Am. Chem. Soc.* **99**, 4803.
148.   A. Camerman, H. W. Smith and N. Camerman. (1976). *Cancer Treat. Rep.* **60**, 517.
149.   H. A. Brassfield, J. C. Clard and J. G. Verkade. (1976). *Cryst. Struct. Comm.* **5**, 417.

150.   A. Camerman, H. W. Smith and N. Camerman. (1975). *Biochem. Biophys. Res. Comm.* **65**, 828.
151.   H. A. Brassfield, R. A. Jacobson and J. G. Verkade. (1975). *J. Am. Chem. Soc.* **97**, 4143.
152.   N. Camerman and A. Camerman. (1973). *J. Am. Chem. Soc.* **95**, 5038.
153.   S. Garcia Blanco and A. Perales. (1972). *Acta Cryst.* **B28**, 2647.
154.   H. Sterngtanz, H. M. Einspahr and C. E. Bugg. (1974). *J. Am. Chem. Soc.* **96**, 4014.
155.   J. C. Clardy, J. A. Mosbo and J. G. Verkade. (1972). *Chem. Comm.* 1163.
156.   J. Durrieu, R. Kraemer and J. Navech. (1972). *Org. Mag. Res.* **4**, 709.
157.   R. Arshinova, R. Kraemer, J.-P. Majoral and J. Navech. (1975). *Org. Mag. Res.* **7**, 309.
158.   C. Roca, R. Kraemer, J.-P. Majoral, J. Navech, J. F. Brault and P. Savignac. (1976). *Org. Mag. Res.* **8**, 407.

7
Seven-Membered and Larger Rings

Introduction

It is obvious that as ring size increases, the degrees of freedom associated with the ring increase and so therefore does the conformational complexity. However, the same rules that we have applied in Chapters 3 to 6 still hold, and the same conformational effects can still be seen in these more complex systems. Amongst the effects that reappear in this chapter are the smaller size of ether oxygen and amine nitrogen than methylene, and hence their preference for sterically hindered regions of a molecule, the *gauche–gauche* preference in the C—O—C—O—C fragment (analogous to the anomeric effect) and the higher torsional potential about bonds between heteroatoms, etc.

It is possible to reduce the conformational complexity of these rings in several ways. The introduction of appropriate substitution patterns such as *gem*-dimethyl groups can raise the energy of certain conformations relative to others, or the introduction of the "small" atoms oxygen or nitrogen can reduce the energy of some conformations relative to others. These changes will also alter the relative energy of transition states for conformational change making analysis of dynamic effects in NMR spectra easier.

Alternatively, the conformational complexity of these larger rings can be reduced by the so-called "rigid group" principle that has been used elegantly and extensively by Ollis's group and many others. In this approach one or more degrees of freedom of the ring are removed by incorporating several

131

of the ring atoms into a rigid planar group. Thus two atoms may be constrained by a double bond or by being *ortho* in a benzene ring. This would,
for example, and as we shall see, effectively give cycloheptene-like systems
the degrees of freedom associated with cyclohexane. Three atoms may be
constrained by their incorporation *meta* in a benzene ring or *peri* in a naphthalene unit, etc.

Because this approach has been used so extensively, only a few of the
more aesthetically pleasing examples are considered here.

The use of *ortho*-disubstituted benzene rings as rigid groups is illustrated
by work on the compounds (1) in which the conformational mobility of the

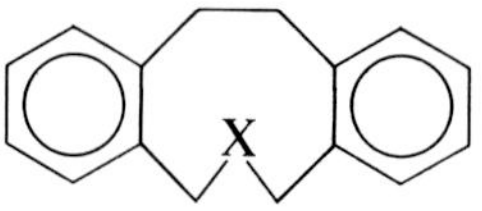

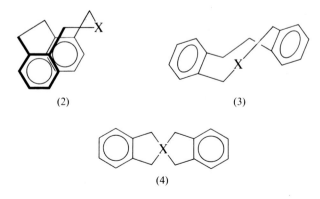

X = NMe, NCH$_2$Ph
S and SO$_2$

(1).

central nine-membered ring is restricted by two such units.[2] At low temperatures the ^{1}H NMR spectra of the methylene groups in the derivatives where
X = NMe, NCH$_2$Ph, S and SO$_2$ are AB quartets requiring that no symmetry
in the molecule should relate the methylene hydrogen atoms, and that therefore the molecules should have either C$_s$ or C$_2$ symmetry. In the spectrum
of the *N*-benzyl derivative the methylene protons of the benzyl group are
also an AB quartet and are therefore also diastereotopic. This observation
eliminates the C$_s$ chair (2) as a possibility, leaving only the C$_2$ chair (3) and the
C$_2$ boat (4) to be considered as candidates. Strain energy calculations finally
allowed a discrimination in favour of the C$_2$ chair conformation (3).

The incorporation of *peri*-annelated naphthalene residues is illustrated by
studies on the dioxonin (5) and the dioxecin (6).[3]

(2)   (3)

(4)

(5)   (6)

(7)

Perhaps the most pleasing work incorporating the rigid group principle is the work on tri-salicylides and related compounds[1,4,5] including tri-o-thymotide (7). Tri-o-thymotide, which spontaneously resolves on crystallization, adopts a propeller conformation of C_3 symmetry in the solid phase. The molecules may also adopt a helical conformation of C_1 symmetry, and a schematic representation of their interconversion is drawn in Fig. 7.1. The inversion of the 12-membered central ring has been studied by variable

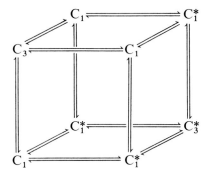

Fig. 7.1   Propeller (C_3) and helix (C_1) interconversions in tri-salicylides such as tri-o-thymotide. Asterisks denote mirror image conformations.

temperature ^{1}H NMR spectra. At low temperatures, lines attributable to both conformations are visible, with the propeller being much the more abundant conformation. As the temperature is raised, two processes with differing activation energies are observed corresponding to the helix → helix interchange and to the propeller → helix interchange.

Racemization of the spontaneously resolved material had been studied earlier by polarimetric methods.[6] It is pleasing, therefore, that both NMR and polarimetry agree very closely on the activation energy for the overall inversion process, (propeller → enantiomeric propeller; 20.5 kcal mol^{-1}, NMR; 20.9 kcal mol^{-1}, polarimetry).

Seven-Membered Rings

The conformations and conformational processes in cycloheptane and its derivatives, and of heterocyclic seven-membered rings are much less well understood than for most of the other simple monocyclic systems. The introduction of the extra methylene group, turning cyclohexane into cyclo- heptane, leads from a system whose conformational properties are well defined and conceptually rather simple to a system that is much less well defined and far more complex. The complexity has, however, been reduced in many seven-membered rings by introducing rigid groups such as o- disubstituted benzene, and by increasing torsional barriers by introducing, for example, di- and tri-thia linkages.

The molecular mechanics calculations of Hendrickson[7] and of Bixon and Lifson[8] on cycloheptane, show the main conformational features to be expected in seven-membered rings. Hendrickson's calculations indicate that the twist–chair conformation of symmetry C_2 (8) is probably the most stable. The twist–chair can pseudorotate very easily crossing a very low energy barrier of a chair arrangement with C_s symmetry (9) to another C_2 twist–chair. This pseudorotation interconverts all substituent positions on the ring, giving, in the case of rapid pseudorotation, a completely time averaged NMR spectrum. Since the calculations suggest a barrier of only ca 2 kcal mol^{-1} for this process in cycloheptane it must occur extremely rapidly even at very low temperatures.

Similar conclusions were reached by Strauss both for cycloheptane and for some related oxepanes (oxygen-containing seven-membered rings), from

(8)   (9)

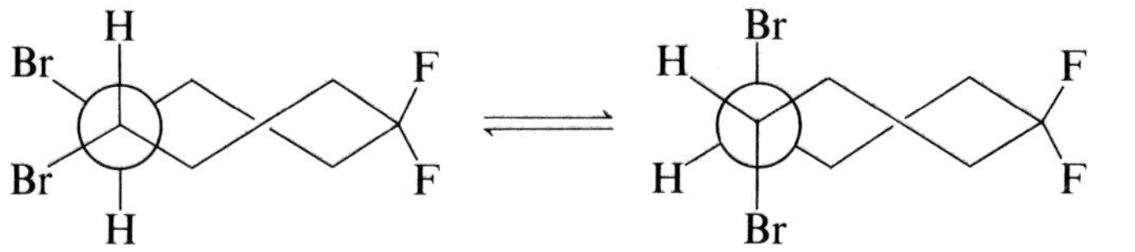

Fig. 7.2   Postulated ring inversion process in *trans*-1,2-dibromo-5,5-difluoro-cycloheptane.

a consideration of vibrational spectral data.[9-11] Spectral assignments,[10] and normal coordinate calculations showed that cycloheptane, oxepane, and 1,3-dioxepane have twist–chair conformations as their lowest energy forms. Cycloheptane and 1,3-dioxepane exist as twist–chairs of C_2 symmetry, whilst oxepane exists as two asymmetric D,L pairs of twist–chairs of approximately equal energy. Subsequent calculations,[11] predicted barriers to pseudorotation of 1.3 and 4.0 kcal mol^{-1} in cycloheptane and 1,3-dioxepane respectively. It is hardly surprising therefore that no NMR evidence can be found for the slowing of pseudorotation or any exchange process in these simple seven-membered rings, even at very low temperatures.[11,12] For example, no sign of any broadening ascribable to an exchange process was observed in the ^{19}F spectrum of 1,1-difluorocycloheptane even as low as $-180°$.

As was suggested above, the pseudorotation and ring inversion processes can be slowed down by the introduction of suitable substituents. Thus, to take first an alicyclic example, the single ^{19}F resonance in the proton decoupled nmr spectrum of (10) at ambient temperature, splits into two singlets in the intensity ratio 74:26 below $-114°$.[13] This result indicates the presence of two conformations in which the fluorines are related by C_s or C_2 symmetry, and was rationalized in terms of a conformational equilibrium between the two twist–chair forms (10a) and (10b) in Fig. 7.2.

A similar effect of heteroatom–heteroatom bonds on rates of conformational interchange is found in the seven-membered hydroxylamine derivative (11).[14] The original authors ruled out nitrogen inversion as the origin of their observations and referred to the process as "conformational inversion".

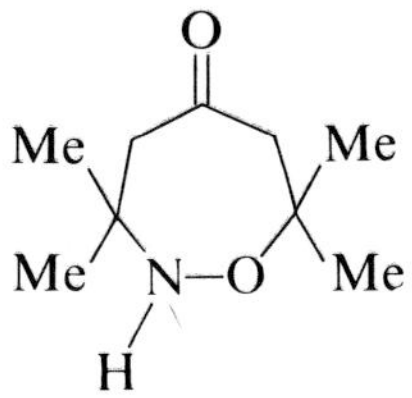

(11)

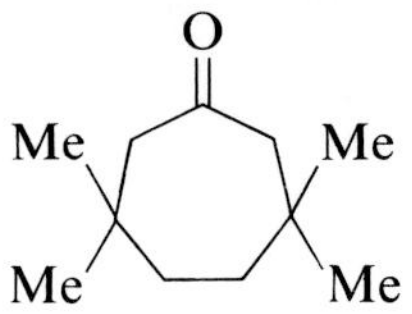

(12)

From a consideration of the results, it seems likely to the present author that a pseudorotation process is being observed. The barrier to this process is 19.2 ± 0.5 kcal mol^{-1} at ca $100°$, and is ca 10 kcal mol^{-1} greater than in the corresponding hydrocarbon (12). This very large increase in activation energy probably arises because the *gem*-dimethyl groups restrict the number of possible conformations on the pseudorotation circuit, and place N—O bond rotation in the rate determining step thus considerably raising the barrier.

The effect of heteroatom–heteroatom bonds is also seen in thia-substituted cycloheptanes.[15,16] For 1,2,3-trithiepane (13) the very low barrier to pseudorotation in cycloheptane is raised to ca 6–7 kcal mol^{-1}.[15] The further introduction of two adjacent sulphur atoms to give 1,2,3,5,6-pentathiepane (14) again increases the observed barrier. The coalescence in the ^{1}H spectrum

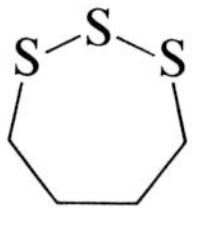

(13) 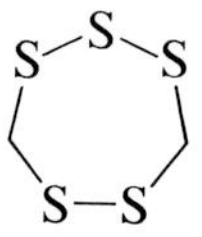   (14)

occurred at $-60°$, and at $-90°$ two singlets were observed for the methylene groups. The absence of any geminal coupling in the methylene resonances strongly suggests that pseudorotation is still occurring to interconvert the environments of the methylene hydrogens. The authors of this work suggest that a ring inversion (14a $\rightleftharpoons$ 14b) is the observed process. However, these conformations form part of the pseudorotation pathway which interconverts

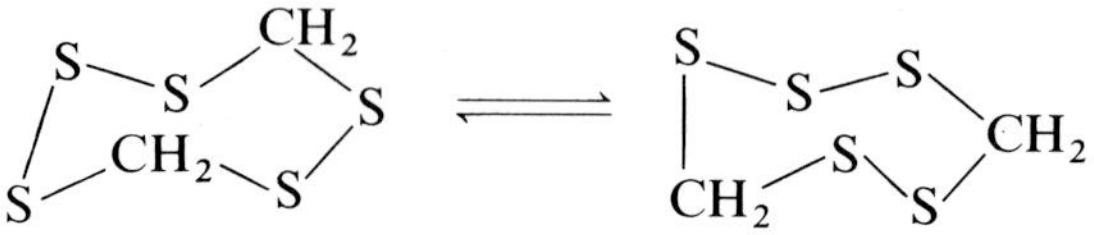

(14a)   (14b)

all environments. Slowing of pseudorotation would give rise to AB quartets for the methylene groups. Thus the observed process may well be a twist-chair $\rightleftharpoons$ twist-boat ring inversion.

The rigidity in the double bond segment of the cycloheptene ring reduces the conformational complexity of this system and brings it into line with

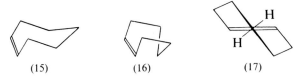

(15) (16) (17)

that of cyclohexane. It is generally assumed that three conformations, the chair (15), the boat (16), and the twist-boat (17) are important. Their inter-conversions are outlined in Fig. 7.3. Again we shall see alterations in the relative energies of ground and transition states brought about by hetero-atom substitution.

Infrared and Raman spectra of solid and of liquid samples of cyclo-heptene are compatible with the molecules being in the chair form in the solid, and largely in the chair form in the liquid phase. Hydrogen-1 NMR results for the benzocycloheptene system strongly favour the chair conforma-tion[18,19] and low-temperature ^{1}H NMR measurements on 4,4,5,6,6-penta-deuteriocycloheptene have been interpreted in terms of a 5.0 kcal mol^{-1} free energy barrier to chair–chair interconversion at $-165°$.[20] An analogous study of 5,5-difluorocycloheptene (18), shows an enthalpy barrier to chair–chair interconversion of 7.4 kcal mol^{-1}.[12]

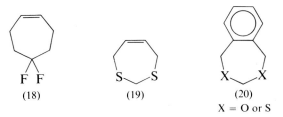

(18) (19) (20)

X = O or S

The dithio analogue of cycloheptene (19), shows similar activation parameters, but the corresponding oxygen-containing ring seems to have an appreciably lower barrier.[21-23] The benzo-systems (20), show barriers that are 2–4 kcal mol^{-1} greater, but again the oxygen-containing rings have lower barriers than those containing sulphur.[21-23] In some of the oxygen-contain-ing rings no barrier was observed. This would be consistent with a change in the preferred ground state conformation from chair to twist-boat making chair–chair interconversion not observable by NMR methods. In accor-dance with this idea, evidence has been produced that *cis*- and *trans*-4,7-dimethyl-1,3-dioxacyclohept-5-enes (21) have a twist-boat conformation.[24]

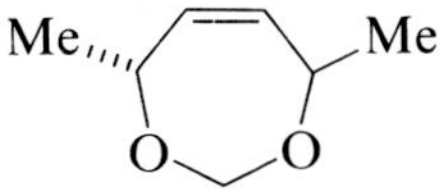

Me,,,,   Me

O   O

cis and *trans*

(21)

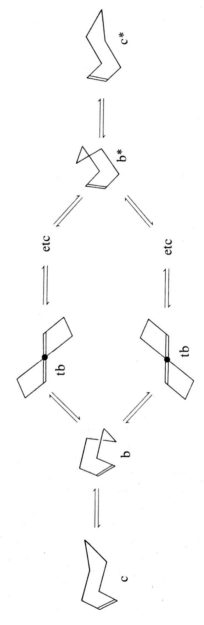

Fig. 7.3   Chair–chair interconversion route in cycloheptene.

In the twisted-boat conformation the oxygen-containing end of the ring may get close to the *gauche–gauche* arrangement of bonds required for C—O—C—O—C bonds by the anomeric effect. A similar preference for a twisted-boat ring is seen in 3-methyl-2-oxo-1,4-dioxepane (22) where the lactone group is approximately planar and is effectively acting as an "ene" group.[25]

Evidence for the coexistence of twist-boat and chair conformations is forthcoming from low-temperature NMR measurements on the benzo-cyclic sulphite (23). The single AB quartet for the methylene group splits and broadens into 2-quartets as the temperature is lowered to $-80°$. It is plausibly suggested that these arise from twist-boat and chair conforma-tions.[26]

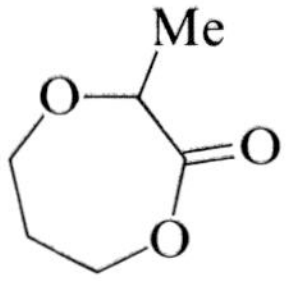

(22) 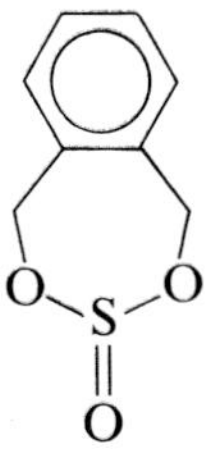   (23) 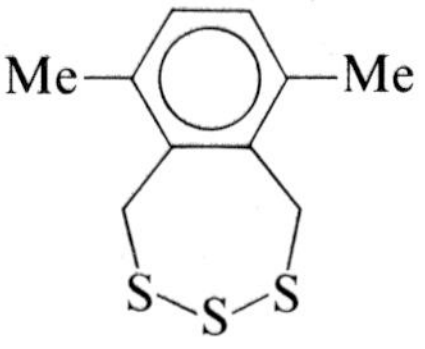   (24)

However, the most convincing piece of work on the ring inversion and pseudorotation processes in heterocyclic cycloheptene analogues is on the tri-thia derivative (24).[27] At $+165°$ its ^1H NMR spectrum consists of three singlets for the aromatic, methyl, and methylene groups. On cooling to room temperature, this pattern changes to two methyl peaks, two aromatic peaks and a singlet and an AB quartet for the methylenes. On further cooling the methylene singlet splits into a second AB quartet. The interpretation of these changes in terms of Fig. 7.3 is straightforward. The high-temperature process is the chair $\rightleftharpoons$ (boat + twist-boat) interconversion ($\Delta G_c^{\ddagger} = 19.8$ kcal mol^{-1}). The methylene singlet is from the pseudorotating twist-boat, and the quartet from the chair. Further cooling slows down the pseudorota-tion splitting the singlet into a quartet ($\Delta G_c^{\ddagger} = 11.5$ kcal mol^{-1}).

Eight-Membered Rings

The more usual experimental methods of conformational analysis have proved to be singularly unrewarding when applied to cyclooctane.[28] The molecule is too complex for the reliable use of infrared and Raman spectro-scopy and the only useful information that these methods give is that the molecules are not centrosymmetric.[29] X-Ray diffraction of crystalline cyclooctane proves of limited use because the crystals are probably disordered.[30] Electron diffraction gives results that cannot be simply interpreted on the basis of any single conformation, and seems to indicate a mixture of conformations in the gas phase, which contain many similar interatomic distances and whose nature could not be further defined.[31] Electron diffraction does, however, show a marked opening of intra-annular valence angles (116°). Earlier NMR studies on cyclooctane and deuterated derivatives only enabled the exclusion of any conformation with a C_2 axis passing through carbon atoms or a C_s plane passing only through bonds.[32]

There are generally two conformations considered to be important for cyclooctane, the crown (25) and the boat–chair (26). Molecular mechanics calculations on cyclooctane are in agreement with the boat–chair conformation of C_s symmetry being slightly more stable than the crown.[7,8,33,34]

(25)   (26)

These calculations reveal that there are many low activation energy ring inversion and pseudorotation processes possible in cyclooctane. These processes have been discussed in some detail by Anet.[28]

More recent NMR results are in accordance with cyclooctane existing as a 95:5 mixture of boat–chair and crown conformations with $\Delta H° = 1.9$ kcal mol^{-1}, $\Delta S = 1 \pm 1$ eu and $\Delta G^{\ddagger} = 11.2$ kcal mol^{-1} at $-45°$ (for boat–chair → crown).[35]

X-Ray diffraction studies on heterocyclic eight-membered rings show that the molecules either adopt crown (27–30)[36–39] or boat–chair (31–33)[40–42] conformations.

Anet and Degen examined the low-temperature ^{1}H NMR spectra of oxocane (34) at high field (251 MHz) and observed a slowing of ring inversion ($T_c = -122°$, $\Delta G_c^{\ddagger} = 7.4$ kcal mol^{-1}).[43] The low-temperature spectra were consistent with the boat–chair conformation (35) in which the intra-annular hydrogen at position 1 has been replaced by a pair of electrons on oxygen, lowering repulsive interaction with the intra-annular hydrogens

(27)   (28)

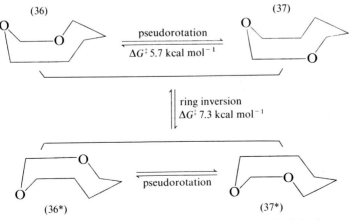

(29)

(30)

(31)

(32)

(33)   (34)   (35)

on positions 4 and 4′. For 1,3-dioxocane, two conformational processes could be frozen out ($\Delta G_c^{\ddagger}$ = 7.3 and 5.7 kcal mol^{-1}). The first process to be frozen out is thought to be ring inversion and the second to be a pseudo-rotation between the boat–chair conformations (36) and (37), Fig. 7.4. These conformations not only allow reduction of non-bonded interactions, but also get the C—O—C—O—C fragment as nearly into a favourable

(36)   (37)

pseudorotation
$\Delta G^{\ddagger}$ 5.7 kcal mol^{-1}

ring inversion
$\Delta G^{\ddagger}$ 7.3 kcal mol^{-1}

pseudorotation

(36*)   (37*)

Fig. 7.4   Pseudorotation and ring inversion processes in 1,3-dioxocane.

gauche–gauche relationship as is practicable in the chair–boat conformation. Both 1,3,6-trioxocane (38) and 1,3,5,7-tetraoxocane (39) show similar results in that each exists as a mixture of two conformations in solution whose interconversions can be frozen out. Similar results were reported by Dale.[44]

For the compound 5-oxocanone both ring inversion ($\Delta G_c^{\ddagger} = 9.0$) and pseudorotation ($\Delta G_c^{\ddagger} = 7.8$ kcal mol^{-1}) can be frozen out and the preferred conformation appears to be (40).[45]

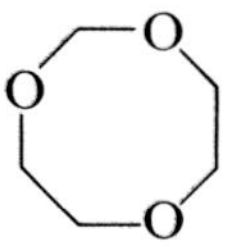

(38) 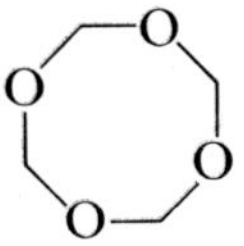   (39) 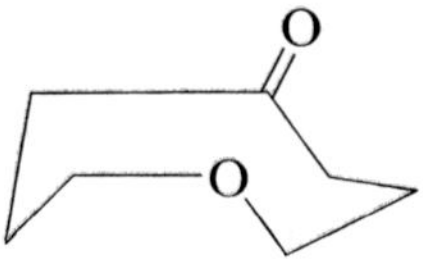   (40)

Anet has also investigated the conformation of azocane by ^{1}H and ^{13}C NMR. The ring inversion barrier of the preferred boat–chair conformation (41) is observed at ca $-120°$ ($\Delta G_c^{\ddagger} = 7.3$ kcal mol^{-1}). A further process is observed in the ^{13}C spectra which arises from interconversion of the boat–chair conformation with ca 3% (at $-112°$) of the crown conformation. ($\Delta G° = 1.2$ kcal mol^{-1}; $\Delta G^{\ddagger} = 10.5$ kcal mol^{-1}).[46] These results are slightly at variance with earlier reported work.[47]

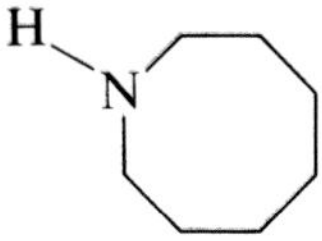

(41)

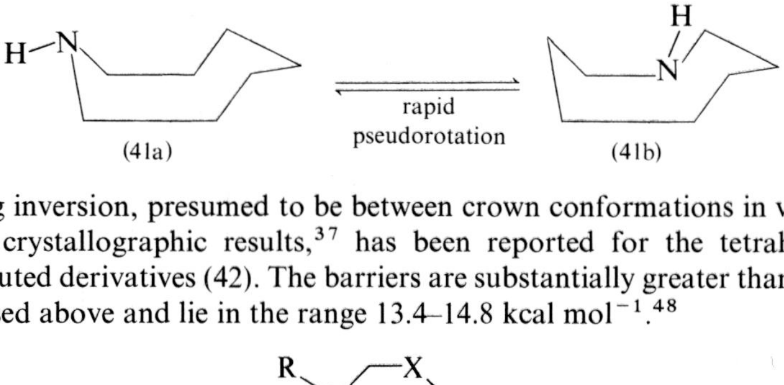

(41a)   rapid pseudorotation   (41b)

Ring inversion, presumed to be between crown conformations in view of X-ray crystallographic results,[37] has been reported for the tetrahetero-substituted derivatives (42). The barriers are substantially greater than those discussed above and lie in the range 13.4–14.8 kcal mol^{-1}.[48]

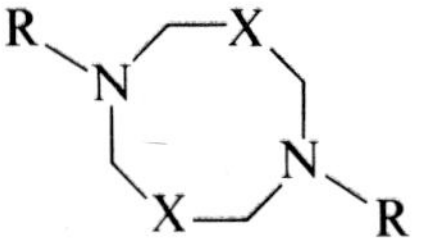

X = S or Se
R = Me, Et, *i*-Pr or Ph
(42)

Nine-Membered Rings

The two main conformations found for nine-membered rings have symmetry C_2 and D_3 and have been called the twist-chair–boat TCB (43) and twist-boat–chair TBC (44) respectively. Trimeric acetone peroxide (45) has been shown by X-ray diffraction to exist in the TBC conformation

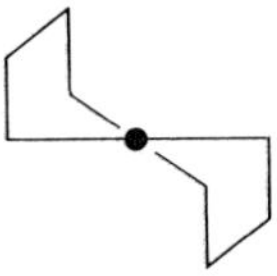

TCB   TBC

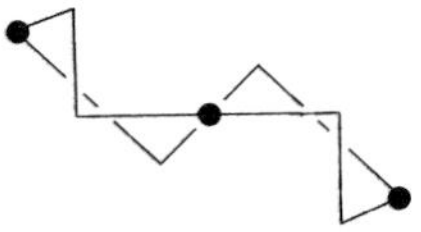

(43)   (44)   (45)

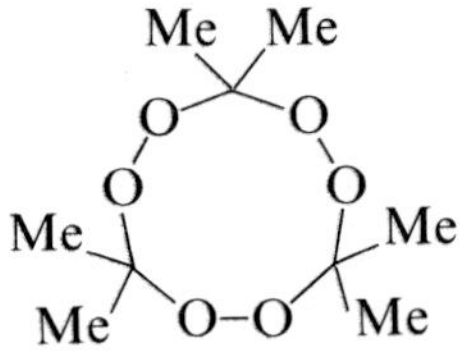

of D_3 symmetry.[49] In this case the presence of three geminal dimethyl groups introduces steric constraints that preclude conformations of lower symmetry being important, and each *gem*-dimethyl group lies on a two-fold axis. On the other hand, cyclononylamine hydrobromide[50] and cyclononanone mercuric chloride[51] have been shown to have conformations of approximately C_2 symmetry, and are therefore twist-chair–boats, possibly as a result of crystal packing forces in these dominantly inorganic crystal lattices.

The low-temperature ^{13}C NMR spectrum of cyclononane at $-162°$ shows two distinct types of carbon atom in the integrated ratio of 2:1.[52] This result is entirely consistent with the D_3 TBC conformation, and is unlikely for any conformation of lower symmetry (e.g. C_2 TCB). Moreover the splitting arises from a conformational process of activation energy ca 6 kcal mol^{-1}, which is in reasonable agreement with calculations on processes that would lead to exchange between the two sites.[53]

As is seen with other ring systems discussed in this chapter the introduction of geminal dimethyl groups can reduce or alternatively reinforce conformational preferences of rings. The preference for the TBC conformations in 1,1-dimethylcyclononane (46),[52] and in 1,1,4,4-tetramethylcyclononane (47)[54] are inferred from NMR data. The presence of the *gem*-dimethyl groups lowers the overall molecular symmetry from D_3 to C_2 in both cases, and the central ring retains approximate D_3 symmetry. The activation

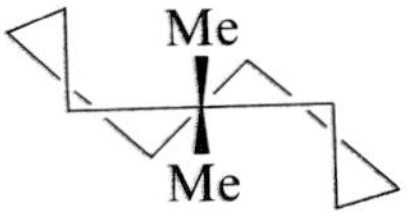

(46) 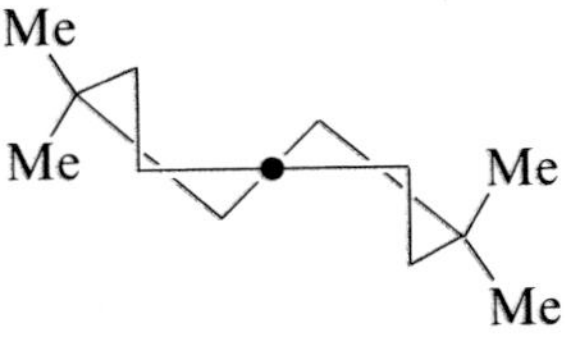   (47)

energies for the observed conformational processes also increase from 9 kcal mol^{-1} (46),[52] to 14–20 kcal mol^{-1} (47 and some derivatives).[54]

Nuclear magnetic resonance spectral data on 3,3,6,6-tetramethyl octane-dioic anhydride (48) are also in complete accord with a TBC conformation in which the anhydride group is symmetrically placed with respect to the two-fold axis.[55]

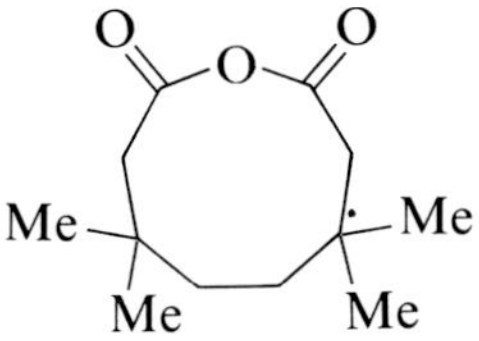

(48) 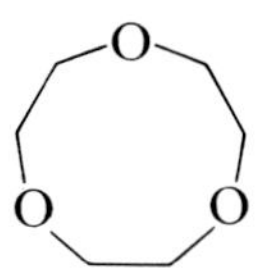   (49)

The trimer of ethylene oxide (49), on the other hand, shows an asymmetrical triangular conformation that is neither a TBC nor a TCB form.[56] The conformation does, however, correspond to a low-energy form predicted in Dale's calculations.[53]

Ten-Membered Rings

The conformational analysis of cyclodecane is fairly well understood largely due to X-ray crystallographic work by Dunitz[57] who showed, by X-ray diffraction, that the preferred conformation of the ring is (50),[58] which has C_{2h} symmetry. In this conformation, which is a distorted part of the diamond lattice, there are three different types of carbon atom. The torsion angles around the ring are approximately staggered and the C—C—C bond angles are opened out to an average value of ca 117°. This bond angle distortion presumably arises because of repulsions between three intra-annular hydrogen atoms on carbons of types I and II which are only 190–200 pm apart. There are fourteen extra-annular hydrogen atoms of four different types all of which are relatively unhindered. Figure 7.5 displays these relationships.

Conformational biasing can be achieved either by the introduction of gem-dimethyl groups, which will avoid intra-annular locations, and therefore locate themselves on carbon atoms of type II, or by the introduction of small heteroatoms which will relieve strain by going to positions I or III. Thus 1,1,4,4- and 1,1,6,6-tetramethyl derivatives are conformationally biassed by the gem-dimethyl groups into conformations depicted in (51) and (52), and 2-oxacyclodecane-1,6-dione (53) shows a conformation with both carbonyl carbons and the ring oxygen atom occupying hindered positions.[59]

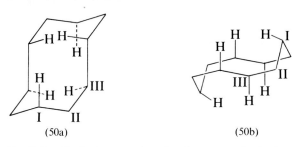

Fig. 7.5 Cyclodecane, showing intra-annular hydrogens.

The "normal" cyclodecane conformation can be disrupted by the special requirements of heteroatoms. Thus the conformation of the cyclic pentamer of formaldehyde (54) found by X-ray crystallography, is completely unlike that of cyclodecane. In it the anomeric *gauche–gauche* requirement of every acetal unit is met. A conformation of this sort would not be expected for

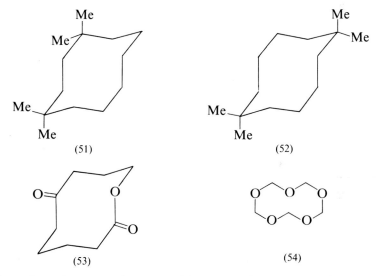

hydrocarbons because of enhanced intra-annular H–H repulsions. These repulsions are relieved by the smaller oxygen atoms in the conformation found for (54).

Larger Rings

With certain specific exceptions very little systematic experimental or computational attention has been paid to rings of eleven or more members.

It is now, however, fairly clear that even-membered rings prefer to adopt conformations which form part of the diamond lattice, and which contain the maximum number of *trans*-butane units.[61] Diamond lattice conformations are not possible for odd-membered rings, whose conformations are therefore almost diamond lattice-like, with a distortion in one segment of the ring, or completely unlike any segment of the diamond lattice. For each ring system containing up to 24 carbon atoms a list is available giving all possible diamond lattice conformations.[62] The conformational consequences of replacing methylene in some of these rings with oxygen have been discussed in some detail, and with an outline of available results by Dale.[63]

Specific systems in this category that have been investigated in some detail include cyclic peptides,[64–66] and the related cyclic oligoesters of glycolic acid.[67] In these systems the amide bonds may adopt E or Z configurations with important consequences for the conformations of the rings.

Crown ethers, cryptates and other macrocyclic complexing agents have also been widely investigated,[63, 68–70] in view of their biological[68] and synthetic applications.

References

1. W. D. Ollis, J. F. Stoddart and I. O. Sutherland. (1974). *Tetrahedron* **30**, 1903.
2. W. D. Ollis and J. F. Stoddart. (1974). *Angew. Chem.* (Int. edn, English) **13**, 730.
3. D. J. Brickwood, W. D. Ollis and J. F. Stoddart. (1974). *Angew. Chem.* (Int. edn, English) **13**, 731.
4. A. P. Downing, W. D. Ollis, I. O. Sutherland, J. Mason and S. F. Mason. (1968). *Chem. Comm.* 329.
5. A. P. Downing, W. D. Ollis and I. O. Sutherland. (1970). *J. Chem. Soc.* (B) 24.
6. A. C. D. Newman and H. M. Powell. (1952). *J. Chem. Soc.* 3747.
7. J. B. Hendrickson. *J. Am. Chem. Soc.* (1961). **83**, 4537; (1963). **85**, 4059; (1964). **86**, 4854; (1967). **89**, 7036, 7043, 7047.
8. M. Bixon and S. Lifson. (1967). *Tetrahedron* **23**, 769.
9. D. F. Bocian, H. M. Pickett, T. C. Rounds and H. L. Strauss. (1975). *J. Am. Chem. Soc.* **97**, 687.
10. D. F. Bocian and H. L. Strauss. (1977). *J. Am. Chem. Soc.* **99**, 2866.
11. D. F. Bocian and H. L. Strauss. (1977). *J. Am. Chem. Soc.* **99**, 2876.
12. J. D. Roberts. (1966). *Chem. Brit.* 529.
13. R. Knorr, C. Ganter and J. D. Roberts. (1967). *Angew. Chem.* (Int. edn. English) **6**, 556.
14. R. E. Wasylishen, K. C. Rice and U. Weiss. (1975). *Canad. J. Chem.* **53**, 414.
15. S. Kabuss, A. Luttringhaus, H. Friebolin and R. Mecke. (1966). *Z. Naturforch.* **21b**, 320.
16. R. M. Moriarty, N. Ishibe, M. Kayser, K. C. Ramey and H. J. Gisler. (1969). *Tetrahedron Lett.* 4883.
17. N. Neto, C. DiLauro and S. Califano. (1970). *Spectrochim. Acta* **26A**, 1489.

18. G. L. Buchanan and J. M. McCrae. (1967). *Tetrahedron* **23**, 279.
19. S. Kabuss, H. G. Schmid, H. Friebolin and W. Faisst. (1969). *Org. Mag. Res.* **1**, 451.
20. M. St. Jacques and C. Vaziri. (1971). *Canad. J. Chem.* **49**, 1256.
21. H. Friebolin, R. Mecke, S. Kabuss and A. Luttringhaus. (1964). *Tetrahedron Lett.* 1929.
22. S. Kabuss, H. Friebolin and H. G. Schmid. (1965). *Tetrahedron Lett.* 469.
23. H. G. Schmid, H. Friebolin, S. Kabuss and R. Mecke. (1966). *Spectrochim. Acta* **22**, 623.
24. M. H. Gianne, M. Adams, H. G. Kuivila and K. Wiersthorn. (1975). *J. Org. Chem.* **40**, 450.
25. P. Ayras and A. Partanen. (1977). *Org. Mag. Res.* **9**, 379.
26. H. Faucher, A. Guimares and J. B. Robert. (1977). *Tetrahedron Lett.* 1743.
27. S. Kabuss, A. Luttringhaus, H. Friebolin, H. G. Schmid and R. Mecke. (1966). *Tetrahedron Lett.* 719.
28. For a review see F. A. L. Anet. (1974). *Fortschritte* **45**, 169.
29. H. E. Bellis and E. J. Slowinski, Jr. (1959). *Spectrochim. Acta* **15**, 1103.
30. D. E. Sands and V. W. Day. (1965). *Acta Cryst.* **19**, 278.
31. A. Almenningen, O. Bastiansen and H. Jensen. (1966). *Acta Chem. Scand.* **20**, 2689.
32. F. A. L. Anet and M. St. Jacques. (1966). *J. Am. Chem. Soc.* **88**, 2585.
33. K. B. Wiberg. (1965). *J. Am. Chem. Soc.* **87**, 1070.
34. N. L. Allinger, J. A. Hirsch, M. Miller, J. J. Tyminski and F. A. van Catledge. (1968). *J. Am. Chem. Soc.* **90**, 1199.
35. F. A. L. Anet and V. J. Basus. (1973). *J. Am. Chem. Soc.* **95**, 4424.
36. L. Pauling and D. L. Carpenter. (1937). *J. Am. Chem. Soc.* **58**, 1274.
37. D. Grandjean and A. Leclaire. (1967). *Compte Rend.* **265**, 795.
38. H. H. Caldy, A. C. Larson and D. T. Comer. (1963). *Acta Cryst.* **16**, 617.
39. H. Schenck. (1971). *Acta Cryst.* **B27**, 185.
40. G. W. Frank, P. J. Degen and F. A. L. Anet. (1972). *J. Am. Chem. Soc.* **94**, 4792.
41. K. T. Go and I. C. Paul. (1965). *Tetrahedron Lett.* 4265; (1969). *J. Chem. Soc.* (*B*). 33.
42. S. M. Johnson, C. A. Maier and I. C. Paul. (1970). *J. Chem. Soc.* (*B*) 1603.
43. F. A. L. Anet and P. J. Degen. (1972). *J. Am. Chem. Soc.* **94**, 1390.
44. J. Dale, J. Ekeland and J. Krane. (1972). *J. Am. Chem. Soc.* **94**, 1389.
45. F. A. L. Anet and P. J. Degen. (1972). *Tetrahedron Lett.* 3613.
46. F. A. L. Anet, P. J. Degen and I. Yavari. (1978). *J. Org. Chem.* **43**, 3021.
47. J. B. Lambert and S. A. Khan. (1975). *J. Org. Chem.* **40**, 369.
48. J. M. Lehn and F. G. Riddell. (1966). *Chem. Comm.* 803.
49. P. Groth. (1969). *Acta Chem. Scand.* **23**, 1311.
50. R. F. Bryan and J. D. Dunitz. (1960). *Helv. Chim. Acta* **43**, 3.
51. S. G. Dahl and P. Groth. (1971). *Acta Chem. Scand.* **25**, 1114.
52. F. A. L. Anet and J. Wagner. (1971). *J. Am. Chem. Soc.* **93**, 5266.
53. J. Dale. (1973). *Acta Chem. Scand.* **27**, 1130.
54. G. Borgen and J. Dale. (1970). *Chem. Comm.* 1105.
55. G. Borden. (1974). *Acta Chem. Scand.* **28**, 13.
56. J. Dale, G. Borgen, F. A. L. Anet and J. Krane. (1974). *Chem. Comm.* 243.
57. For a review see, J. D. Dunitz. (1968). *In* "Perspectives in Structural Chemistry", Vol. 2, (Eds Dunitz and Ibers), Wiley, New York.
58. See for example, E. Huber-Buser and J. D. Dunitz. (1960). *Helv. Chim. Acta* **43**, 760.
59. I. J. Borowitz and G. Gonis. (1964). *Tetrahedron Lett.* 1151.

60. Y. Chatani and K. Kitahama. (1973). *Bull. Chem. Soc. Jpn* **46**, 2300.
61. J. Dale. (1963). *J. Chem. Soc.* 93.
62. M. Saunders. (1967). *Tetrahedron* **23**, 2105.
63. J. Dale. (1974). *Tetrahedron* **30**, 1683.
64. Yu A. Ovchinnikov and V. T. Ivanov. *Tetrahedron* (1974). **30**, 1871; (1975). **31**, 2177.
65. J. Dale, P. Groth and K. Titlestad. (1977). *Acta Chem. Scand.* **B31**, 523.
66. J. Dale and K. Titlestad. (1978). *Tetrahedron Lett.* 379.
67. J. Dale, O. Sevaldsen and K. Titlestad. (1978). *Acta Chem. Scand.* **B32**, 306.
68. W. Burgermeister and R. Winkler-Oswatitsch. (1977). *Fortschritte* **69**, 93.
69. H. E. Simmons, C. H. Park, R. T. Ugeda and M. F. Habibi. (1970). *Trans. N.Y. Acad. Sci.* **32**, 521.
70. J. M. Lehn. (1973). *Struct. Bonding (Berlin)* **16**, 1.

Index

5-Alkylidene-1,3-dioxans, 72
Anet equations, 37
Anomeric effect, 29, 68, 79, 90, 95, 110, 116, 117, 118, 123, 131, 141.
    see also Reverse anomeric effect, 30
Axial-equatorial energy differences, 24, 25
  in cyclohexane (Table), 19
  in alkylpiperidines, 87
  in 1,3-dioxans, 72–76
  in 1,3,2-dioxaphosphorinans, 123
  in 1,3-dithians, 109
  in 1,3,5-dithiazines, 117
  in hexahydropyrimidines, 90
  in hexahydro-1,3,5-triazines, 92
  in *N*-methylpiperidine, 86
  in 1,3-oxathians, 113, 114
  in 1,4-oxathians, 117
  in phosphorinans, 120
  in piperazines, 91
  in tetrahydro-1,4,2-dioxazines, 97
  in tetrahydro-1,2,4-oxadiazines, 97
  in tetrahydro-1,2,5-oxadiazines, 98
  in tetrahydro-1,2-oxazines, 94
  in tetrahydropyrans, 69
  in selenan-1-oxide, 107
  in 1,3,5-thiadiazines, 117
  in thian derivatives, 105, 106
1-Aza-3-thiabicyclo[4,4,0]decane, 117
Azetidine, 28, 53
Azocane, 142

Bohlmann bands (infrared), 42
Bond angle deformation, 4, 7, 12, 13
Bond lengths, 22–26
Bond rotation
  carbon–carbon bonds, 28, 29
  carbon–heteroatom bonds, 28
  heteroatom–heteroatom bonds, 26–28, 70, 88, 93, 123, 131
  table of barriers, 15
Bond stretching, 7, 12
n-Butane, 8, 9
Butyrolactone, 58

Chair–twist energy differences, 25, 26
  in 1,3-dioxan, 73, 77, 78
  in 1,3-dithian, 109
  in 1,3-oxathian, 115
  in 1,2,4,5-tetrathian, 112
  in thian, 105
Charge alternation effect, 84
Conformation (definition), 3–6
Conformational analysis (definition), 3
Conformational trapping, 33, 86, 87
  in cyclohexane, 33
  in piperidine, 33, 34, 86, 87, 91
Crown ethers, 146
Cryptates, 146
Cyclic voltammetry, 89
Cyclobutane derivatives, 28, 47–50
  axial and equatorial substituents, 49

Cyclohexane, 1, 2, 17, 18, 33
  axial substitution, 18
  boat, 1
  chair, 1, 10
  chair–twist energy difference, 10, 26
  equatorial substitution, 18
  twist, 10, 18
Cyclohexene, 40
Cyclopentane, 55
  envelope conformation, 55
  half-chair conformation, 55
  ring-puckering pseudorotation, 55
Cyclopentene, 61
Cyclic peptides, 146
Cyclophosphoramide, 125
Cytidylic acids, 56

Decahydroquinoline derivatives, 99, 100
Decalin, 1
1,3,2-Diazaphosphorinan derivatives, 125
Dihydro-1,3,5-dithiazine derivatives,
  117, 118
Dihydro-1,4,2,3-dithiadiazines, 118
2,3-Dihydrofuran, 61
2,5-Dihydrofuran, 61
Dihydro-1,2-oxazine derivatives, 38
1,3-Dioxacycloheptene derivatives, 137
1,2-Dioxan derivatives, 70
1,3-Dioxan derivatives, 32, 38, 40,
  70–78
  axial–equatorial energy differences,
    72–76
  chair–twist energy difference, 26, 77, 78
  equilibration reactions, 72
1,4-Dioxan derivatives, 68, 69, 78–80
1,3,2-Dioxaphosphorinan derivatives,
  122–124
  with three ligands on P, 123, 124
  with four ligands on P, 124, 125
1,3,2-Dioxathian derivatives, 27, 117
Dioxene, 42, 70
1,4-Dioxepane derivatives, 135, 139
1,3-Dioxocane derivatives, 141
1,3-Dioxolan, 56, 57, 58
  derivatives of, 57, 58
1,4-Diphosphorinan derivatives, 121
1,2-Diselenan, 107
1,4-Diselenan, 111
1,2-Dithian derivatives, 107, 108

1,3-Dithian derivatives, 33, 38, 108–110
  axial–equatorial energy differences,
    109
  chair-twist energy difference, 26, 109
  equilibration reactions, 108, 109
1,4-Dithian derivatives, 110, 111
1,3,2-Dithiaphosphorinan derivatives,
  125
1,3-Dithietane derivatives, 54
1,3-Dithiolan, 58, 59
Dipole moment measurements, 2, 40, 41,
  57, 85, 90, 95, 96, 118
  relation to molecular symmetry, 40
Dynamic nuclear magnetic resonance,
  35–37

Eight-membered rings, 139–142
Electron diffraction, 3, 39, 53, 56, 58, 78,
  80, 140
Electrostatic interactions, 7, 13, 14, 29,
  30, 76
  charge–charge, 13
  charge–dipole, 13, 76
  dipole–dipole, 14
  octupole interactions, 14
  quadrupole interactions, 14
Epimerization, 32, 72
Equilibration reactions, 2, 32, 33
  in 1,3-dioxans, 32, 72–77
  in 1,3-dioxolans, 58
  in 1,3-dithians, 108, 109
  in 1,3-dithiolans, 59
  in 1,2-oxaphosphorinans, 121
  in 1,3-oxathians, 113
  in 1,3-oxathiolans, 59
  in phosphorinans, 120
  in thianium salts, 105
Ethane, 8, 15, 16
Ethylene sulphate, 59

Far infrared spectra, 42, 51, 56, 59
Five-membered rings, 54–61
Four-membered rings, 47–54

Gauche effect, 28, 67, 73, 75, 79
Gauche repulsive effect, 29, 69, 76, 79,
  110, 116, 117
Glucuronolactone, 58

Hexahydropyridazine derivatives, 87–89
Hexahydropyrimidine derivatives, 89–91
Hexahydro-1,2,4,5-tetrazine derivatives,
    43, 92
Hexahydro-1,3,5-triazine derivatives, 92
Hockey sticks effect, 29, *see also Gauche
    repulsive effect*
Hydrazine, 15, 26
Hydrazine derivatives, 43, 87–89, 92,
    93, 98, 99
Hydrogen bonding, 7, 41, 42, 75, 76
Hydrogen peroxide, 15, 26
Hydrogen peroxide derivatives, 27, 70
Hydroxylamine, 15, 26
Hydroxylamine derivatives, 60, 93, 96,
    97, 98, 135
5-Hydroxymethyl-1,3-dioxans, 76

Infrared spectra, 33, 34, 41–43, 47, 52,
    86, 91, 93, 95, 96, 118, 121, 137,
    139

Karplus equation, 34, 35, 57
Kerr constants, 85

Macrocycles, 146
2-Methoxy-1,3-dioxans, 76
Mass spectral appearance potentials, 78
Microcalorimetry, 77, 78
Microwave spectroscopy, 39, 47, 50, 51,
    52, 58, 67, 80, 86
Molecular geometry, 23
Molecular mechanics, 7, 59, 67, 105, 134,
    140
Molecular orbital (MO) calculations,
    6, 8, 30
Molecular polarizability, 85
Molecular rotation measurements, 77

Nicotine, 60
Nitrogen inversion, 5, 37, 60, 83, 84, 88,
    89, 92, 93, 95, 97, 98
Nine-membered rings, 132, 133, 143, 144
Non-bonded interactions, 7, 11, 12,
    22–26, 73, 94, 95, 131
    attractive, 107, 123
    exponential function for , 11

Non-bonded interactions—*cont.*
    Lennard–Jones function for, 11, 12, 23
Nuclear magnetic resonance spectra,
    34–39, 77, 95, 108, 113, 118
    ^{13}C, 38, 88, 90, 91, 97, 106, 109, 142,
        143
    coupling constants, 34, 35, 96, 106,
        113, 116, 119, 125
    ^{19}F, 135
    ^{1}H, 38, 85, 88, 90, 93, 97, 106, 111,
        112, 125, 132, 136, 137, 139, 142
    ^{31}P, 120, 125
    shift reagents, 37, 38
    variable temperature, 35–37, 86, 88,
        92, 134, 143,
Nucleosides, 56

1,3,4-Oxadiazolidine derivatives, 60
1,2-Oxaphosphorinan derivatives, 121
1,3-Oxaselenan derivatives, 115
1,2-Oxathian derivatives, 113
1,3-Oxathian derivatives, 33, 113–115
1,4-Oxathian derivatives, 68, 69, 116,
    117
1,3-Oxathiolan derivatives, 59
Oxaphosphetane derivatives, 53
1,3,2-Oxaphosphorinan derivatives, 125
Oxazetidine derivatives, 54
1,2-Oxazolidine derivatives, 60
1,3-Oxazolidine derivatives, 60
Oxepane derivatives, 134, 135
Oxetane, 28, 29, 50, 51
Oxocane derivatives, 140–142

1,3,5,7,9-Pentaoxacyclodecane, 145
1,2,3,4,5-Pentathian, 112
1,2,3,5,6-Pentathiepane, 136
Phosphetane derivatives, 53
Phosphorinan derivatives, 119–121
Phospholane derivatives, 60, 61
Phosphorus-containing six-membered
    rings, 118–125
Photoelectron spectra, 43, 89
Piperazine derivatives, 91
Piperidine, 39, 42, 83
    and derivatives, 31, 32, 40, 83–87
Piperidinium salts, 29
Polarimetry, 134
Proline, 60

Pseudolibration, 56, 78
Pseudorotation, 18, 55, 56, 58, 78,
  135–139
Pyranose sugars, 31, 66, 68
Pyrazolidine derivatives, 60
Pyrrolidine derivatives, 60
Pyrrolizidine derivatives, 60

Quinolizidine derivatives, 99

Raman spectra, 1, 41–43, 47–49, 137, 139
Resonance effects, 7, 87
Reverse anomeric effect, 30
Ribose, 56
Rigid group principle, 131–134
Ring dihedral angles, 28, 35, 71, 78, 79,
  93, 106, 107, 108, 119
Ring inversion, 5, 28
  in cyclobutane, 48–50
  in 1,3-dioxan, 78
  in 1,4-dioxan, 79, 80
  in 1,3,2-dioxathian, 117
  in 1,2-dithians, 107, 108
  in eight-membered rings, 140–142
  in hexahydropyridazines, 88
  in hexahydropyrimidines, 89
  in hexahydrotriazines, 36, 92
  in 1,3-oxathian, 115
  in piperazines, 91
  in selenan, 67
  in seven-membered rings, 135–139
  in telluran, 67
  in tetrahydro-1,4,2-dioxazines, 97
  in tetrahydro-1,3-oxazines, 95
  in tetrahydro-1,4-oxazines, 96
  in tetrahydropyran, 67
  in thians, 104, 106
R value, 35, 67, 108, 111
Rotamer, 6

Selenan derivatives, 86
Seven-membered rings, 134–139
Silacyclobutane, 52
Solvent effects, 7, 73, 75
Stereoisomers, 3
Sulphur groups, 52, 54, 106, 108, 109,
  110, 111, 113, 117

Sulphur-containing six-membered rings,
  104–118

Telluran derivatives, 86
Ten-membered rings, 133, 144, 145
Tetrahydrofuran, 56
Tetrahydro-1,2,4-dioxazine derivatives,
  97
Tetrahydro-1,2,5-dioxazine derivatives,
  98
Tetrahydro-1,4,2-dioxazine derivatives,
  96, 97
Tetrahydro-1,3,4-oxadiazine derivatives,
  98, 99
Tetrahydro-1,2-oxazine derivatives,
  93–95
Tetrahydro-1,3-oxazine derivatives, 95,
  96
Tetrahydro-1,4-oxazine (morpholine)
  derivatives, 85, 96
Tetrahydropyran derivatives, 29, 30,
  67–70
  oxonium salts of, 68
Tetrahydroselenophene, 39
Tetrahydro-1,3,5-thiadiazine derivatives,
  117
Tetrahydro-1,3-thiazine, 117
Tetrahydro-1,4-thiazine derivatives, 118
Tetrahydrothiophene, 39
1,4,5,8-Tetraoxadecalin, 79
1,2,4,5-Tetraoxan derivatives, 24, 80
1,3,5,7-Tetraoxocane, 142
1,2,4,5-Tetrathian derivatives, 111, 112
Thian derivatives, 25, 86, 104–107
Thianium salts, 105
1,3-Thiazolidine derivatives, 60
Thietane, 51
Thiophane, 58
Torsion angles, see Ring dihedral
  angles
Torsional forces, 7–11, 26–28, 70, 88, 89,
  93, 123, 131
Trimethylene selenide, 52
Trimethylene sulphite, 117
Tri-o-thymotide, 133, 134
1,3,5-Trioxan, 80
1,3,6-Trioxocane, 142
1,2,4-Trioxolan derivatives, 58
Tri-salicylides, 133
1,2,3-Trithian derivatives, 111

1,3,5-Trithian derivatives, 109, 111
1,2,3-Trithiepane, 136
Twelve-membered rings, 133, 146
Twist conformations
    in cyclohexane, 4, 10, 18, 26, 34
    in 1,3-dioxan, 26, 77, 78
    in 1,3-dithian, 26, 109
    in 1,3-oxathian, 115
    in 1,2,4,5-tetrathian, 111, 112
    in thian, 105

Ultrasonic relaxation, 39, 40, 78

Van der Waals radius, 18, 22–26, 83, 94,
        104, 131
    table of, 23

X-ray diffraction
    of azetidine derivatives, 53
    of cyclohexane derivatives, 2
    of 1,3-dioxan derivatives, 71

X-ray diffraction—*cont.*
    of 1,4-dioxan derivatives, 79
    of 1,3,2-dioxaphosphorinan
        derivatives, 122
    of 1,4-diphosphorinan derivatives, 121
    of 1,2-dithian derivatives, 107
    of 1,3-dithian derivatives, 108
    of 1,4-dithian derivatives, 110
    of 1,3-dithietane derivatives, 54
    of 1,3-dithiolan derivatives, 59
    of eight-membered rings, 139–142
    of hexahydropyridazine derivatives, 88
    of 1,4-oxathian derivatives, 116
    of 1,3,2-oxazaphosphorinan
        derivatives, 125
    of phosphorinan derivatives, 119
    of piperazine derivatives, 91
    of piperidine derivatives, 84
    of ten-membered rings, 144, 145
    of tetrahydro-1,2-oxazine derivatives,
        93, 94
    of thian derivatives, 105
    of thietane derivatives, 52
    of 1,3,5-trioxan derivatives, 80

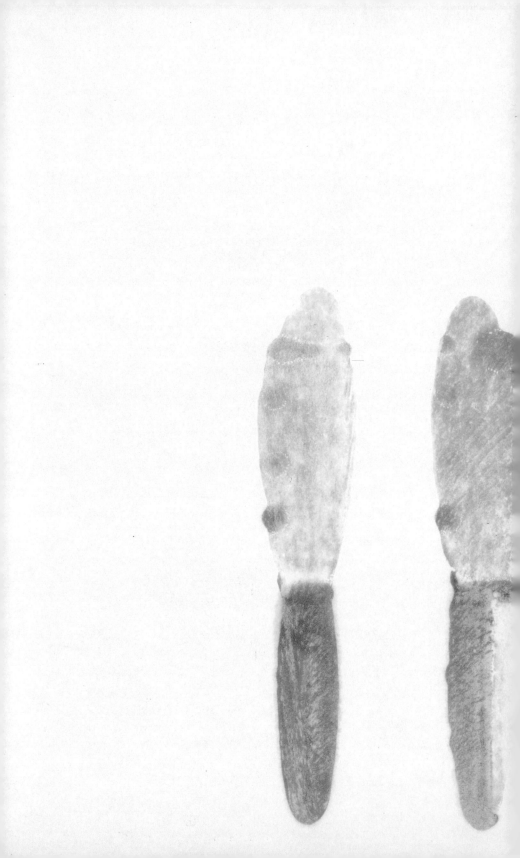